T0211489

Astronomers' Universe

Series Editor
Martin Beech
Campion College
The University of Regina
Regina, Saskatchewan, Canada

The Astronomers' Universe Series is aimed at the same people as the *Practical Astronomy Series*—in general, active amateur astronomers. However, it is also appropriate to a wider audience of astronomically-informed readers.

Because optical astronomy is a science that is rather at the mercy of the weather, all amateur astronomers inevitably have periods when observing is impossible. At such times they tend to read books about astronomy and related subjects.

When researching this market, it is quite surprising to discover how few books there are that are of direct appeal to "armchair astronomers". There are many "popular science" books about matters cosmological, but because of their general audience these all start from the beginning, covering and re-covering the basics. At the other end of the spectrum there are professional books that are highly mathematical and technical, not intended for enjoyable reading.

The *Astronomers' Universe Series* begins by assuming an appropriate level of knowledge. Basic information about the distance, the solar system, galaxies, etc. is not part of these books, which can take a basic understanding of this as their starting point.

The series is differentiated from popular science series (such as Springer's *Copernicus* books) by a strong design image which will attract active amateur astronomers, and will also appeal to "armchair astronomers" (or cosmologists) and other readers who already have the necessary background knowledge.

The books have to be carefully written, structured and edited so as to be aimed at these scientifically-aware readers: they will have a background knowledge of astronomy and probably cosmology but many of them will not have formally studied science (amateur astronomers come from almost all walks of life) and will be discouraged by mathematical treatments. The content will therefore mostly be descriptive, with only essential mathematics included.

More information about this series at http://www.springer.com/series/6960

Karim A. Malik • David R. Matravers

How Cosmologists Explain the Universe to Friends and Family

 Springer

Karim A. Malik
School of Physics and Astronomy
Queen Mary University of London
London, UK

David R. Matravers
Institute of Cosmology and Gravitation
University of Portsmouth
Portsmouth, UK

ISSN 1614-659X   ISSN 2197-6651   (electronic)
Astronomers' Universe
ISBN 978-3-030-32733-0   ISBN 978-3-030-32734-7   (eBook)
https://doi.org/10.1007/978-3-030-32734-7

To our parents.

Preface

Discussing our work over the years with friends, family members, and other non-cosmologists, we frequently get asked "what is it you do *exactly*" to which we usually say "well, we study cosmology." The conversation then usually takes one of two turns: either we give a single sentence explanation along the lines of "we study the universe and its origins," in which case our counterpart often feels short changed and rightly so, or we try to explain in some detail, to pay justice to the topic we spend our lives researching, but our counterpart gets overwhelmed by the rather complicated subject and loses interest.

Both outcomes were and are really frustrating for us and our friends. We don't want to describe our work and main research interest with a short sentence, but we also don't want to lecture our friends and watch their eyes glaze over when talking about equations and observational data.

We would have liked to recommend a nontechnical book, and there are many commendable ones, but we could find none that would explain in sufficient detail while at the same time being accessible to a person without a Mathematics or Physics background. The books we saw available were either too simplistic, making things unnecessarily mysterious, or too complicated, expecting too much prior knowledge.

This book aims at filling this gap. We started 6 years ago, thinking "how hard can it be?", but it turned out to be quite difficult. To explain modern cosmology without using equations, and hence make it accessible for somebody without a physics background, but without oversimplifying the subject, is rather difficult.

This book aims to provide an up-to-date overview of modern cosmology, in particular the evolution of the large-scale structure, the distribution of galaxies and clusters of galaxies, in the universe. Cosmology is a complex area of research, without question. But nevertheless, it is well within the grasp of a "layperson," if this person is prepared to keep an open mind and happy to spend some time letting some of the concepts sink in. To help the reader, we try to explain most technical concepts within the book and also added a glossary at the end where we define and briefly explain the most frequently occurring technical terms. The book is aimed at the interested layperson with little or no physics background but an interest in modern cosmology.

In the brief introduction, we try and give a rounded, complete picture of our current understanding of cosmology based on the latest research and in particular the structure of the universe on very, very large scales. We hope that this is not too overwhelming but instead entices the reader to find out more! Besides, we have to start somewhere. We are aware that therefore the introduction will be rather challenging, but we ask the reader to persist, as most or all of the topics will then be discussed in detail and put into context.

After the introduction, we provide the reader with an overview of the scientific method, how observations are made and what kind of observations are there. We then discuss the constituents of the universe, including dark matter and dark energy, and provide an overview of the forces shaping the universe, in particular gravity. We thus take the reader on a tour back in time from the present day to the very beginning and discuss the beginning of the universe, a period called inflation, which sets the scene, or the initial conditions, for the following evolution. We end with a concluding chapter.

London, UK   Karim A. Malik
Westbourne, UK   David R. Matravers
August 2019

Contents

Acronyms

Throughout this book, we try to avoid the use of acronyms where possible. However, sometimes, acronyms have become the common name for an experiment or phenomenon, and we can't avoid using the acronym.

2dF	Two-Degree-Field Galaxy Redshift Survey
CDM	Cold Dark Matter
CMB	Cosmic Microwave Background
COBE	Cosmic Background Explorer
ELT	Extremely Large Telescope
ESA	European Space Agency
ESO	European Southern Observatory
Gpc	Gigaparsec
HST	Hubble Space Telescope
IR	Infrared
JWST	James Webb Space Telescope
kpc	Kiloparsec
LIGO	Laser Interferometer Gravitational-Wave Observatory
LOFAR	Low-Frequency Array
LSS	Large-Scale Structure
MACHO	Massive Astrophysical Compact Halo Object
Mpc	Megaparsec
pc	Parsec
SDSS	Sloan Digital Sky Survey
SKA	Square Kilometre Array
VLA	Very Large Array

VLT	Very Large Telescope
WIMP	Weakly Interacting Massive Particle
WMAP	Wilkinson Microwave Anisotropy Probe

1

Introduction

The world is everything that is the case.
Wittgenstein

When we look up into the sky at night we see—not a lot if we live in a city, maybe the moon and some bright stars plus a planet or two. However, if we leave the bright city and its street lights behind, the picture changes dramatically. We see thousands of stars, and after watching a while we can discern patterns: we might notice that the stars are not evenly distributed. We might see that there is a band of stars right across the sky, this is our galaxy. If we have binoculars to aid our star gazing, we might notice that some of the dots we thought were stars are indeed tiny blobs, little "nebulae" or galaxies like our own. Using more sophisticated and bigger telescopes reveals even more structure in the distribution of galaxies on the largest visible scales. We are still learning more about the universe, as astronomers, with more sophisticated equipment, continue to make new observations at wavelengths beyond those that are visible to the naked eye, for instance observing in the radio and microwave wavelengths.

But how did the structure in the distribution of galaxies we observe form? How did the universe evolve to look as it does today, and how did it all start? These are some of the questions that cosmologists try to answer, and which we discuss in this book. We will deal with physical cosmology which can be regarded as the scientific study of the origin, formation and evolution of structures on large scales and their dynamics, and the ultimate fate of

© Springer Nature Switzerland AG 2019
K. A. Malik, D. R. Matravers, *How Cosmologists Explain
the Universe to Friends and Family*, Astronomers' Universe,
https://doi.org/10.1007/978-3-030-32734-7_1

the universe, as well as the scientific laws that govern the various processes involved. We will also discuss the science that is needed to make sense of all the observational data that astronomers provide.

In this book we aim to give a non-technical overview of the current observational and theoretical developments and understanding in cosmology, with a slight emphasis on the theory side, as both authors are theorists. The main problem or difficulty we faced in writing a non-technical book, is that the only precise and exact way to represent the inner workings of theoretical cosmology is in the form of mathematics and equations. Cosmology is part of modern physics and the language of physics, its essence, is mathematics.

In fact much of the mathematics used in cosmology is hard even for the experts, and therefore even harder on the lay-person. We therefore decided to leave all equations and formulae out of the main body of the text, only having some minimal set of example equations in the appendix, where they can't do much harm. However, there is a limit to what we can do with words only, as the essence of the physical processes is in the equations. We have to own up to this and therefore, on occasion, we have to refer to "equations" when "just using words" would not be enough to clarify things or worse, would mystify what is really going on and why cosmologists came up with some otherwise inexplicable results. Where appropriate we will also use drawings, figures, tables or diagrams to clarify or illuminate the ideas.

Remarkably, most of the physics underlying the evolution of the universe is the same or at least very similar to the physics we resort to everyday to, for example, generate electricity, or design and build cars and fridges. Only during extreme epochs in the history of the universe, such as the very beginning of the universe, and at very strange places, such as close to a black hole, does the standard physics break down and we have to resort to exotic physics, luckily for both cosmologists and the reader. We will describe the relevant physics, without equations, later on when needed.

1.1   The Cosmological Standard Model

Let us begin with an outline of the history of the universe according to the *cosmological standard model*, which describes what most cosmologists might agree upon as to how the universe evolved, under the tacit understanding that nothing "too weird" is included. We should point out that this cosmological standard model is not as broadly accepted as the standard model in particle physics,[1] and many, or even most, aspects of it are as yet unclear or unknown. But it is this that makes the subject so exciting because there is still so much to understand and so much being revealed by researchers even today. This picture of the cosmos was derived in the last 100 years from our understanding of the fundamental physics and it is constantly revised in accordance with the observations and experiments, and theoretical developments.

In this chapter we will only give a very brief overview of the main ideas in cosmology, and then we will explain the physics and fill in the details and explanations in the chapters that follow. We are aware that the reader will be confronted with many new terms and concepts, however, we hope that the reader will eventually understand what the rough sketch of the "history of the universe" presented in this chapter is all about by reading on. Most of the technical terms introduced in this chapter are also discussed in the glossary at the end of the book. But we are confident that the reader will be able to read through this and the following chapters without relying too much on the glossary, and the technical terms will become clear in later chapters in context.

Recent years have seen a remarkable transformation of cosmology, from a rather esoteric subject at the borderline of theoretical physics and applied mathematics to one of the most vibrant and popular areas of modern astronomy. This development has been brought about by the spectacular advances on the theoretical side of cosmology, but more importantly, by the availability of new telescopes and other sophisticated equipment to gather observational data of a quantity and quality that was unthinkable just a couple of decades ago.

Figure 1.1 summarises the evolution of the universe as described in the cosmological standard model, and highlights some of the crucial events in its history. The different epochs are discussed in the following sections. The universe has a beginning, which is subject to intense speculation and the physical and mathematical machinery available is still inadequate to say much or anything useful about "time zero". Hence we shall exclude this singular point in time from the discussion for now.[2]

---

[1]The "standard model of particle physics" is the theoretical model that explains the properties and interactions of the known subatomic particles, in excellent agreement with the experimental evidence, and therefore agreed upon by the scientific community.

[2]Some authors use "time zero" or "Big Bang" to denote the beginning of the universe, but we find both terms not very useful since they might lead to confusion. We therefore shall simply talk about "the beginning" when we mean the beginning of the universe.

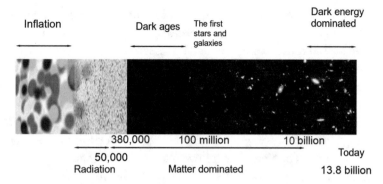

Fig. 1.1 The evolution of the universe in what has become the cosmological standard model. Some key events and epochs in the history of the universe are highlighted (time in years). The evolution of the universe ended being dominated by radiation 50,000 years after the beginning, when matter began to dominate. This epoch lasted until the universe was roughly 10 billion years old. The Cosmic Microwave Background formed after 380,000 years, the first stars and galaxies after 100 million years. Today the universe is 13.8 billion years old

1.1.1 From Inflation to Radiation Domination

At the beginning of the universe, the earliest epoch that we can say something definite about, is a period of extremely rapid expansion of space called "inflation", in Fig. 1.1 on the very left. Although inflation only lasts for the tiniest fraction of a second, it nevertheless manages to inflate—hence the name—the universe by many orders of magnitude, distances get stretched by a huge factor. This epoch starts at roughly 10^{-36} s after the very beginning.[3] During inflation small fluctuations in the density of the "particle soup" present at this very early time, are stretched by the extremely rapid expansion of space to comparatively very large scales, and they become "frozen in", in other words they become stuck and stop fluctuating. The scales or distances over which these fluctuations are stretched are much larger than the distance light, or anything else, could have travelled in the time since the beginning of the universe.

These primordial, generated during inflation, density fluctuations act as "seeds" for the later evolution of structure in the radiation and matter, and throughout the history of the universe these small fluctuation are further amplified through the "gravitational instability": regions that have a slightly higher density than their surroundings attract more material and hence their

---

[3]See Appendix A.1 for an explanation of small and large numbers.

density grows. This means they will attract even more material, further enhancing the density of the region and so on. Through these processes the primordial density patterns become imprinted on the temperature fluctuations of the Cosmic Microwave Background (this is often abbreviated as "CMB"[4]) and the Large Scale Structure (often abbreviated as "LSS"), the distribution of matter in the universe today on the scales of hundreds of millions of light years and more.

The traces of these primordial patterns in the sky are used by cosmologists to discover a great deal about how the universe evolved from very early times. The universe keeps expanding from the beginning to the present day, however the speed of its expansion changes over time. An essential feature of inflation is that the expansion is not only extremely rapid but that the expansion speed increases, that is the expansion accelerates. In later epochs the expansion speed of the universe slows down, but in cosmologically recent times, the expansion started to speed up again. More details and descriptions of how all this works will be given in the following chapters, in particular Chap. 6 on the forces that govern the universe with a particular focus on gravity, and a discussion of the gravitational instability in Sect. 6.4.3, Chap. 7 on how the structures on the largest scales formed, Sects. 7.4 and 8.2.3 on the Cosmic Microwave Background, and Chap. 9 which discusses inflation.

1.1.2 From Radiation to Matter Domination

The particle responsible for, or "driving", inflation conveniently decays at the end of this epoch into standard matter constituents and radiation, or photons.[5] At this point in time the temperature is so high, that all the particles present form a sort of photon-particle-soup. The temperature is even too high for protons and neutrons to form. After inflation ends, at about 10^{-34} s after the beginning, the universe keeps expanding, but at a much slower rate than before, and once the temperature has dropped enough, particles more familiar to us, like the proton and the neutron, can form. As the temperature decreases further these particles then bind together to form the nuclei of

---

[4]Throughout this book we try to avoid not only scientific, technical jargon, but also the use of acronyms where possible.

[5]By "radiation" we mean electromagnetic waves, and the particle associated with electromagnetic waves is the photon. When referring to "matter" we usually have in mind both normal matter (the particles that form atoms), and the exotic dark matter (which makes its presence felt only gravitationally). We will discuss electromagnetic radiation in more detail in Sect. 3.2.1, the constituents of the universe in Chap. 5, and in particular dark matter in Sect. 5.4.

the lightest elements, such as hydrogen, helium, and lithium during what is known as "nucleosynthesis", which takes place about 100 s into the evolution. At this temperature "standard nuclear physics" takes over, albeit standard physics applied to matter and radiation in extremis. The universe is still in an incredibly hot and dense state at these early times, but cools down as it further expands. However, as we already pointed out, the expansion is much more gentle than during inflation. From the end of inflation, when the particles governing this period decay into standard matter, until roughly 50,000 years after the beginning the matter content of the universe and its evolution are dominated by the radiation or photons. This period is therefore also known as the "radiation dominated era", see Fig. 1.1.

1.1.3 From Matter Domination to the Present Day

By the time the universe was roughly 380,000 years old it had cooled sufficiently, to about 3000 K, for photons to stop interacting with other particles, in particular electrons, see Fig. 1.1. This event is called "decoupling", because the standard matter decouples from the radiation, that is the photons. These photons, which form the Cosmic Microwave Background, were left to travel forward in time towards us without interference, with the patterns of its interaction with matter at decoupling preserved. When the radiation reaches us today its temperature is about 3 K and it is in the form of microwave radiation.

Because the evolution of the universe after radiation domination is determined by the matter content and radiation only plays a minor role, this epoch is called the "matter dominated era", and lasts from roughly 50,000 years after the beginning for about 10 billion years. Decoupling takes place early on in this epoch.

As can be seen in Fig. 1.1, decoupling is followed by an epoch known as the "dark ages" during which no objects emitting visible light are present. The first generation of stars are formed roughly 100 million years afterwards, and the universe begins to brighten up again. These first stars are much more massive than the later generations, around a hundred times the mass of the Sun, and their lifetimes much shorter, of the order of tens of millions to hundreds of millions of years (a star like the Sun has a lifetime of the order tens of billions of years). After they have spent all their fuel they collapse under their own gravity, and explode as supernovae. In these violent explosions the first heavy elements, for example carbon and oxygen, are formed and spread across the universe. The next generation of stars then forms from the debris of the earliest stars and the leftover gas from the early universe. The first galaxies begin to assemble during the dark ages, and ever larger structures such as clusters and super-clusters of galaxies form thereafter. For example our own galaxy, the Milky Way, formed when the universe was about 1 billion years old, and it took another 8 billion years, or until 4.6 billion years ago, until the Sun and the solar system formed from the debris of previous stars.

Then, to the surprise of cosmologists when they discovered it at the end of the twentieth century, the universe began another period of accelerated expansion a few billion years ago. This is because roughly 10 billion years after the beginning, or 3.8 billion years ago, another mysterious contribution to the total energy budget, dark energy, takes over and begins to dominate the evolution of the universe. This is the epoch we find ourselves in at the moment, 13.8 billion years after the beginning. The time-line of key events in the history of the universe is summed up in Fig. 1.1, and in more detail in Table 8.2, together with the temperature of the universe at these times.

1.1.4 What Constitutes the Universe

Throughout its history the universe expanded, extremely rapidly for a short time during inflation, and at a more moderate pace later on during radiation and matter domination and recently the expansion has begun to accelerate again. The rate of expansion, how fast the universe expands, is determined by the energy constituent or constituents at a particular time; this will be discussed further in Sect. 6.4.4. The temperature of the radiation in the universe is also related to the expansion: as the universe expands the temperature decreases. We will return to the expansion of the universe in Sect. 6.4.2.4, discussing the expansion-temperature relation in Sect. 6.4.4.

The contribution of the main constituents to the overall or total energy density[6] of the universe at the present day, according to observations updated recently by the Planck satellite, are roughly 68% dark energy, 27% dark matter, and 4.8% baryonic matter, this is shown on the right panel of Fig. 1.2. "Baryonic matter", which for cosmologists includes protons, neutrons, and electrons, is the well known matter that makes up the things we can see in the sky and on Earth. Radiation contributes only at the sub-percent level to the total energy today. "Dark energy" and "dark matter" are labels given to

---

[6]We use "energy density" synonymously with the probably more familiar term "mass density", which denotes the amount of material per volume. This is not very rigorous, but common practice in cosmology, as energy and mass are closely related, as we will discuss in Sect. 6.4.2.1.

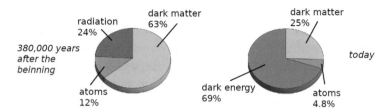

Fig. 1.2 The constituents of the universe, then and now: the "pie" on the left shows the energy or matter content when the universe was 380,000 years old, on the right today, 13.8 billion years later. The numbers indicate the contributions of the different types of "stuff" in percent to the overall energy budget. Early on: the universe contained 63% dark matter (light blue), 24% radiation (red), and 12% of standard matter or atoms (green). Today the dark matter only contributes 25% to the total energy budget, normal matter 4.8%, and dark energy 69% (dark blue). Today radiation only contributes at the sub-percent level and therefore doesn't show up in the diagram (the same holds for the dark energy at early times)

quantities which play key roles in the history of the universe. They are known through their gravitational effects but no other interactions between these dark constituents and other matter have been found, making them difficult to analyse and study. Because of their important role in cosmology, we will discuss them in greater detail in Chap. 5.

The composition of the energy content was different at earlier times, because radiation, matter and dark energy respond differently to the expansion of the universe, each component dilutes at a particular rate, which we will discuss in detail in Sect. 6.4.4. Therefore 380,000 years after the beginning dark matter contributed 63% to the total energy budget, and radiation 24%. The fraction of "normal" matter is 12%, and the contribution of dark energy is at the sub-percent level. This is shown on the right panel of Fig. 1.2. We picked this particular point in time, because this is when the photons we observe today as the Cosmic Microwave Background started their journey, becoming a major source of information about the early universe. From Fig. 1.2 we also see what we mean by an epoch of the universe being dominated by a particular constituent: although there are other contributions to the total energy budget, it is convenient to name an epoch in the history of the universe by the component that is largest. Hence, when the universe was 380,000 years old it is "matter dominated", and today its evolution is dominated by dark energy. As mentioned already, the expansion rate is very different depending on which component in the total energy budget is dominant, a point we will return to in Sect. 6.4.4.

We should stress again that unlike the "standard model" in particle physics, the cosmological standard model is by no means accepted by the whole community working in the wider field of cosmology. This reflects, on the one hand, the heterogeneity and diversity of the field and the community and, on the other, the still weaker experimental and observational "underpinnings" of cosmology compared to particle physics, despite the huge progress in recent years. The cosmological standard model is however the model most practitioners will agree upon, albeit some with slight hesitation.

Above we sketched very roughly how in the cosmological standard model we explain the structure of the universe on large scales and its evolution, how the standard particles and light elements are formed early on during nucleosynthesis, and heavy elements later on in supernovae explosions. But we also mentioned that this model has many as yet hypothetical ingredients, the price we have to pay for the model to work, namely a period of inflation at the earliest times of the universe, and dark matter and dark energy, dominating the recent epochs in the history of the universe.

1.2 Two Pictures of the Universe

We will discuss in detail how we picture the universe and what it looks like on different length scales in Chaps. 3 and 4. But it is useful to present here already two "snapshots" of how the universe looked, taken early on in the history of the universe and taken today, both on the very largest scales.

At early times the Cosmic Microwave Background presents us with a snapshot of what the universe looked like 380,000 years after the beginning. At this time the universe was filled mainly by dark matter and a 3000 K hot plasma[7] with fairly uniform density. But the distribution of the dark matter and the plasma in the universe were not entirely smooth, there were regions that were a little bit more dense and other regions where the material was slightly less dense than the average. These small variations or fluctuations in density also corresponded to fluctuations in the temperature of the plasma: overall slightly denser regions are hotter, less dense regions colder. Figure 1.3 shows these tiny fluctuations in the density and temperature of the plasma. The coloured regions in the figure are hotter or colder by just one part in 100,000,

---

[7]A plasma is one of the fundamental states of matter: the atoms in a gas lose some of their electrons (usually, at standard conditions, all electrons in an atom are bound to the nucleus). We will discuss this in Sect. 5.1.2.

Fig. 1.3 Pictures of the universe: temperature fluctuations in the Cosmic Microwave Background as measured by the Planck satellite. The coloured areas correspond to slightly hotter and colder regions in the Cosmic Microwave Background, of the order of 10^{-5} or 1 part in 100,000 around the average temperature of 2.73 K. All of the sky, the celestial sphere, is mapped onto the ellipse (similar to carefully peeling a satsuma in a single piece and flattening out the peel, in this case into an ellipse). *Image: ESA/Planck*

compared to the average. We will discuss the Cosmic Microwave Background in detail in Sect. 8.2.3.

The oval shape of the map of the Cosmic Microwave Background is a product of the map making. We observe or perceive the Cosmic Microwave Background sky as a giant sphere surrounding us. This sphere is then projected onto the ellipse, similar to carefully peeling an orange or a satsuma, and stretching the peel into elliptical shape—or put more scientifically, using a "Mollweide projection".

The electromagnetic radiation that was emitted by the hot and cold regions gets stretched on its journey to us due to the expansion of space which also implies that the temperature decreases, and we observe this radiation today as microwave radiation with an average temperature of just 2.73 K. The Cosmic Microwave Background today is still very smooth, but luckily for us the small temperature fluctuations of one part in 100,000 are still there for us to measure. The distribution and size of these fluctuations provides us today with invaluable information about the universe at very early times, when it was just 380,000 years old.

At late times, that is today or 13.8 billion years after the beginning, the universe looks very different to the early times. As already mentioned, the universe keeps expanding and cooling. In the time between the two snapshots the universe has expanded by roughly a factor of 1100—distances got stretched 1100 times, volumes increased more than a billion times—and the temperature

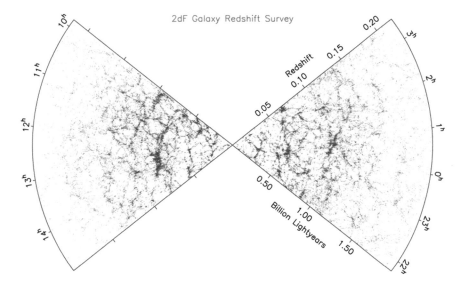

Fig. 1.4 Pictures of the universe: the distribution of galaxies as seen by the galaxy survey "2dF". Each individual dot corresponds to a galaxy. We, the observers, are at the centre of the two cones looking out. We see that there is structure in the distribution of galaxies, up to a maximum size of a couple of hundred million lightyears, but there are no larger structures. Beyond roughly 2 billion light years, the galaxies are too far away and therefore too faint to be observable by the telescope used in the survey. *Image: The 2dF Galaxy Redshift Survey*

decreased by the same factor.[8] The small fluctuations in the density of the dark matter and the plasma present in the early universe grow through gravitational attraction as the universe evolves. In particular, the regions that are slightly more dense than their surroundings attract more material than less dense regions and become even more dense.[9] In these overdense regions galaxies form, and today we can map the distributions of these galaxies in the universe.

Figure 1.4 shows the observed distribution of galaxies today. The image shows two thin, conical slices of sky, with our own galaxy at the centre of the cones, each individual dot corresponds to a galaxy. We can see that on small scales, by cosmological standards, there is structure in the distribution of galaxies, there are filaments, the distribution appears lumpy. But on larger

---

[8] The expansion only affects the largest scales, as we will discuss in Sect. 6.4.2.4, the temperature we referred to is that of the Cosmic Microwave radiation.

[9] We will discuss the gravitational instability—the runaway effect starting from small "overdense" or "more dense than average" regions—in Sect. 6.4.3 and the formation of structure on large scales, that is the distribution of galaxies and dark matter, in general in Chap. 7.

scales, beyond a few 100 million parsecs or a couple of billion lightyears,[10] the distribution of galaxies is featureless and smooth, and there are no larger structures beyond these scales. In addition to the distance in lightyears (at the bottom of the slice), Fig. 1.4 also gives the distance to the galaxies in terms of "redshift". Whereas lightyears or parsecs are just a different distance unit like metres—albeit much, much larger—redshift, how much the light from a galaxy gets stretched by the expansion of the universe on its way to our telescopes on Earth, is not immediately recognisable as a useful measure of distance. However, as we will see in Sect. 6.4.2.4, it can be directly related to distance once we know the laws of physics that underlie the expansion of the universe. The reason redshift is used in cosmology is that it is a directly measurable quantity, like the direction towards the galaxy (the numbers on the sides of the cones are "hour angles", 1 h being equivalent to 15°). The actual distance, be it in metres or lightyears, has then to be calculated.

In both Figs. 1.3 and 1.4 we see that the universe is smooth on the very largest scales. Figure 1.3 is a snapshot of the universe when it was just 380,000 years old. It shows the universe had only tiny fluctuations in its density, at the level of 1 in 100,000, when the Cosmic Microwave Background photons began their journey. These photons are observable today as fluctuations in the temperature of the Cosmic Microwave Background radiation. Figure 1.4, a map of the distribution of galaxies in the universe today, shows that the universe is still smooth on the very largest scales and there is no structure on scales larger than a couple of hundred million parsecs. However on scales smaller than this there clearly is structure recognisable in the figure: filaments along which the galaxies are lined up, surrounding large empty areas or voids, forming what cosmologists refer to as the "cosmic web". Explaining how these structures formed between the time Fig. 1.3 "was taken" at the time the Cosmic Microwave Background formed, and 13.8 billion years later, that is today, is one of the great success stories of cosmology and will be discussed in some detail in Chap. 6 where we look at the forces shaping the universe and Chap. 7 where we explain how these forces lead to the structures we observe in the universe. We would like to emphasise that both of these images are more than simple snapshots. They required the work of hundreds of scientists and represent the latest research in cosmology.

But how did cosmologists come up with this picture of the universe and its evolution, usually referred to as the cosmological standard model? More

---

[10] A "lightyear" is the distance travelled by light in 1 year, a "parsec" corresponds roughly to a distance of 3 lightyears. The distance units used in astronomy are discussed in Sect. 4.1.

importantly what forces and processes governed this evolution? Answering these questions will make up most of the remainder of this book.

As pointed out above, luckily most of the physics that we need to understand most of the evolution of the universe, is "fairly" straight forward, and is the same physics that we encounter in our everyday lives, but that doesn't mean it is trivial. There are two exceptions when exotic physics comes into play; one occurs at the very beginning of the universe and the other in the rather late universe, both require periods of accelerated expansion.

We think it surprising—and fascinating—that so "little" weird and exotic physics is needed to explain most of the evolution of the universe, when it could require some highly abstract and complicated theory. Of course it may do in the end when the physics required to understand the beginning of the universe is resolved.

Although both authors are theorists, we feel compelled by the observational successes made in the last couple of decades—Figs. 1.3 and 1.4 are just two examples of these breakthroughs—to put at the beginning of this book two chapters on observations. In Chap. 3 we will briefly introduce the tools that astronomers use to make these observations and then present and discuss further examples in Chap. 4. In the following chapter on the "stuff" the universe is made of and the forces that govern the universe, Chap. 5, we will proceed from the mundane to the exotic: from the familiar matter that we are made of to the stuff that only cosmologists deal with. In Chap. 6 we discuss the forces that shape the universe, with particular focus on gravity, as this is the force closest to heart of most cosmologists. We describe in some detail how structure forms in the universe in Chap. 7. We then have all the ingredients to take a journey back in time from the present day to inflation, a fraction of a second after the beginning, in Chap. 8, which uses the contents of all previous chapters. In the penultimate chapter of the book we will describe the exotic physics of the most recent era together with the physics of the earliest epoch, the inflationary period, as they exhibit many similarities. We end the book with a brief concluding chapter of a more speculative nature.

Before we can move on and explain what observations are relevant for cosmology and what physics governs which epoch of the universe and on what scales, we need to take a closer look at the scientific method, or how science itself works. The way it works is by no means obvious as we shall see.

2

How Does Science Work?

In this chapter we will try and answer the, at first, very innocent looking question, how does science actually work? In particular, how did we come up with the science that is "physics"? This will take us to the definition of the scientific method, that is the interplay of theory and experiment, which in cosmology, like all other areas of science, is far from trivial.

2.1 Introduction

In Chap. 1 we described the evolution and properties of the universe as they are understood today according to the cosmological standard model, briefly summing up the knowledge we have of the universe. The essence of the model was already named and being worked on in the 1970s. It was derived using the theory, ideas, models and methods of physics which provide the tools to study electromagnetism, gravity, mechanics, fluid flow and other "physical" phenomena that we encounter every day in the world around us. More abstractly speaking, physics is the science that aims to provide a theory for the general analysis and functioning of the physical world.

An aim of the rest of this book is to show how the cosmological standard model was constructed and to explain at least some of the physics behind it. We start with an explanation of how science works, in other words we will have a closer look at the process and methods by which the *theory of physics* is constructed. Inevitably we will briefly touch on the history and philosophy of science and its method but our intention here as in other parts of the book

© Springer Nature Switzerland AG 2019
K. A. Malik, D. R. Matravers, *How Cosmologists Explain
the Universe to Friends and Family*, Astronomers' Universe,
https://doi.org/10.1007/978-3-030-32734-7_2

is not to follow those topics further than we need to and so we will provide references where appropriate to enable readers to read more on the topics if they wish to do so.

To begin our investigation in how science works we will first discuss the scientific method which forms the basis for carrying out science. We will then go on to discuss two simple examples of the applications of the method. The second example will lead us then to a discussion of some of the long held beliefs about science. The chapter ends with a discussion of the increasingly important role of simulations using computers in physics and cosmological science.

2.2   The Scientific Method

Physics, like all the sciences, relies on the so-called scientific method. By this name we mean a procedure which is used to formulate principles, laws or models to describe scientific phenomena. In the ideal case this works as follows:

- We start out with a hypothesis, usually a hunch or prejudice, and further targeted observations and experiments then suggest a possible scientific law or principle to be formulated.
- The next stage in the process is for the hypothesis or law to be rigorously tested by experiments or observations. The experiments are usually specifically designed to test particular predictions based on the provisional law in an attempt to falsify the law. Alternatively, we can confront the hypothesis with observations that have to be consistent with it.
- If it survives, and continues to survive further tests, then the provisional law becomes part of scientific theory.

Note that at any future date a law may be negated by an experiment or observation which contradicts it. Thus, in a sense, all science is, in principle, provisional.[1] The scientific theory is that which survives the experimental tests

---

[1]Some people, especially some social scientists, emphasise a social or personal element which is omitted in the "scientific method" they argue that dominant figures drive the research and can influence the tests and the questions asked. Also they can influence the interpretations and the so bias the answers and conclusions and thus what goes into the theories constructed around the laws. While this is undoubtedly true at times and for individuals or groups, the effect is temporary because of the way the scientific method works.

over time. In physics the laws are usually formulated as concise mathematical statements (laws) which express properties of the physical system.

It is also worth keeping in mind, that hypotheses and physical laws can only be proven wrong or ruled out, never proved right. This is very nicely illustrated by an example introduced by the Austrian philosopher Karl Popper[2] using a black swan: in this case our hypothesis based on observations is that all swans are white. This cannot be proved, because we can never rule out that there might be a black swan somewhere, that we simply haven't "discovered" yet. And indeed, the discovery of black swans has ruled out the original theory. However, these things usually never stop some people from "knowing" that there are no black swans.

As this example shows, putting theories to the test is a crucial part of the scientific method. However as time goes on we can have increasing confidence in the reliability of predictions based on our theory but still a scientific theory or law cannot be proved.

The black swan example above also nicely illustrates the close relation between observations and experiments. Although in the above example the experimental setup was such that observing a non-white swan would rule out our theory, we could have modified our prescription for the scientific method. Instead of performing the experiment "looking-for-swans" we could just have stated our theory that until proven otherwise, there are only white swans.

Indeed we will find that taking observational data, for example observing the distribution of galaxies in the sky, often takes the place of experiments in cosmology. The scientific method requires our theory to be consistent with data in general, whether it came from a dedicated experiment or observations.

Empirically testing, that is not through dedicated experiments, but by requiring that the theory is consistent with observational data is also essential in other areas of physics. Of increasing importance in this context are also simulations run on large computers, which we will discuss in more detail below.

There is, however, much more to the scientific method than discovering laws or formulating theories. But for our purposes we do not require such an in-depth study. The essential feature is that the laws and theory should be tested by using them to formulate hypotheses, which may include alternative laws, that can be tested by experiment. The totality of these laws and or models that apply to the physical world of matter, motion, heat, fluids, electromagnetism,

---

[2]Karl Popper (1902–1994), Austrian philosopher and psychologist, major contributions to the philosophy of science.

atoms, and elementary particles and the theories built upon them, once tested, is what we call physics. While this description is correct, as far as it goes, the process is a little more flexible and complicated as can be seen in the second of the examples that follow.

So far we have tacitly assumed that "laws" that "govern" nature and the physical world exist. This is by no means obvious or guaranteed (by whom we might be tempted to ask). The terminology chosen, "law", expresses the hope or the belief that there is at least some permanence to these governing laws. We might even hope, that they are immutable, even eternal which seems quite daring considering how long some human laws keep. But then we would hope that the laws of nature are more fundamental than the edicts governments issue.

Unfortunately we cannot test whether these physical laws that scientists try to discover, and then understand and apply, are indeed immutable. We can only use the scientific method in the present, indeed there might be a black swan coming along at any moment, destroying our theory. However, luckily, at least we have some knowledge of the past, and we can therefore often be confident that a particular law has been as it is now for the time we have observational data or records to confirm it.

That physical laws exist is therefore an assumption. It is a good assumption in our opinion, since life in its current form would be difficult to develop without continuity in the laws that govern the environment, that is physical laws. In a totally random universe, it is hard to envisage life developing, let alone life that then ponders about the absence of physical laws.

So far we have not properly defined what we mean by a "physical law" or a "law of nature". Indeed most readers of this book already have a good idea what we mean by it. For us here it usually is a basic set of rules and fundamental principles, encoded in the language of mathematics in a set of equations; mathematics also provides the means to solve these equations. This set of equations will also contain at least some fundamental constants of nature, such as π, relating the circumference of a circle to its diameter, or G, Newton's gravitational constant specifying the strength of gravity itself. We can then use this physical law, in form of the equations, to tells us something about the physical world.

The choice of what is a constant of nature and what is not is part of the theory. For example, in some recently popular theories of gravity Newton's constant was assumed to be varying with time (it was "promoted" to be a "field"). The scientific method will also enable us to decide this question: if the observational data or an experiment cannot be explained with a particular quantity being constant, then the theory has to be adjusted. Hence, even constants might change.

We also have to decide *where* we expect our physical laws to hold. It is usually tacitly assumed that physical laws are the same everywhere. This is of particular importance for astronomy and cosmology, since we are applying the laws derived here on Earth to more or less all of space. However since there is no reason why we should assume that things are special on Earth, scientist are quite confident that the laws of physics that apply here are valid everywhere. This concept of universal sameness is referred to as the Copernican principle: there are no special places. We will encounter this principle later again in Sect. 6.4.2.4.

Unfortunately, we can only directly perform experiments on Earth or in the solar system, and even then we are limited by the difficulties of space flight and the amount of money at our disposal. There is no scientific or physical reason that prevents us from undertaking experiments away from Earth, this is simply for financial reasons, as sending satellites into space is rather expensive. However, observations allow us to access other regions. We can observe physical processes in operation elsewhere in the universe without leaving Earth and study whether the outcome of the processes is in agreement, that is consistent, with our expectations. This is not the same as performing experiments elsewhere. But it gives us the opportunity to check that our theories make predictions that are consistent with the observational data from different parts of the universe.

But what if we have two competing theories, that both describe the physical phenomenon we want to describe equally well? In this case we can invoke *Occam's razor*, which states that if we are faced with two competing theories, that both agree equally with the observational data, we should choose the theory with the fewest assumptions, that is the simpler theory.

Finally, all of the above is frequently ignored by scientists. Often they will follow a hunch, which is just a nice way of describing a personal preference or prejudice about how they *think* the world and their area of science *should* be. Or their sense of aesthetics, the beauty of simplicity or symmetry, will guide them, even force them to ignore the data, at least for a while. This topic is closely related to the question how theory changes in physics, which we will discuss in a bit in Sect. 2.4. Let us first discuss two examples, on how the scientific method works.

2.3   Examples

To illustrate the above ideas on how the scientific method works, we will use two simple examples, both related to gravity. The first example is, arguably, not very realistic, nevertheless if taken seriously it is surprisingly difficult to refute.

2.3.1 Colour Dependent Gravity

For the first example of such an experiment let us assume, that we have devised a theory that predicts that red coloured objects fall more slowly than objects of a different colour. This seems far fetched but let us stick with it for the sake of argument.

The experimental setup here would be to take two balls, one red and one white, say. In order to make the experiment relevant to the theory, we would have to carefully control the setup: make sure that the balls have the same mass, that all other influences such as friction are excluded. But let us assume the experiment is exactly as it should be and it only tests the hypothesis, namely that the red ball falls slower.

We perform the experiment, and, to our disappointment, as far as we can measure both balls fall with exactly the same speed. From this experiment we therefore deduce that our theory is wrong, gravity is colour-blind.

If this had been a "real" theory, we wouldn't have given up so easily. We might have assumed that the setup of the experiment was not careful enough, for example that we used balls of slightly different size, or of differently pure material (and hence different mass). We could also have tweaked our theory, done some extra calculations and predicted that the effect is very small, indeed too small to measure. And we would hope for a better experiment, done more carefully or running a much lengthier series of experiments, in the near future.

If we are lucky, we might have found a better theory, where, for example, red objects in the solar system behave differently to red objects further away and only the further away ones obey the slow-fall rule. This would make it much more difficult to rule out or disprove our new theory. However, we would then have to come up with a very good reason why our theory predicts, and indeed needs, this kind of "conspiracy". Our colleagues might indeed invoke Occam's razor, and point out that the theory is now too complicated for no obvious gain, and refuse to look further for evidence for or against our theory. We might give it another try and look for evidence outside the solar system, using telescopes. But having spent some considerable amount of telescope time, we would still come up empty handed. Now we might want to lobby the research council for some money to build a bigger dedicated telescope that …

However, eventually, we would have to abandon our theory. The observational data, in this case from our specific experiments undertaken on Earth and from not observing anything that might support our theory anywhere at all, was just too overwhelming.

By the way, as we will see in Sect. 6.4.2 we should point out that the example wasn't actually as silly as it might sound. It is not obvious that all things should couple in the same way to gravity, or put another way, that there is only one "gravitational charge".

2.3.2 The Perihelion Shift in the Orbit of Mercury

Let us now look at a real example, again involving gravity. The development of the theory of gravity at the beginning of the twentieth century illustrates how the process works or has to be modified to continue to fit the observational constraints. Newtonian[3] theory of gravity was seen as incredibly successful and until the mid nineteenth century when measurements were becoming more precise it still seemed complete. Then, in 1859, it was found that when the orbit of Mercury was studied in detail Newtonian gravity did not give the correct value for the size of the small deviation in the orbit called the perihelion precession, or the shift of the perihelion (see Fig. 2.1). Precession arises because the orbits of the planets about the Sun are not perfect circles, they are elliptical. These ellipses have their nearest and furthest points from the Sun at their turning points of the ellipse. The turning point nearest the Sun is called the aphelion and the furthest point is called the perihelion. When the perihelion shift of Mercury was measured it was found that after corrections for the effects of the other planets on the orbit of Mercury and for the effect of the fact that the Sun is not a perfect sphere as it is slightly flattened there was still a small discrepancy. This discrepancy remained until 1915 when Einstein's[4] general relativistic theory of gravity was shown to provide the required small extra correction needed.

More recent measurements and calculations of the corrections give for the effect (per century): the other planets cause a shift of approximately 531 arc

---

[3]Issac Newton (1642–1727), English physicist and mathematician, ground breaking work in many areas of of physics, one of the developers of calculus.

[4]Albert Einstein (1879–1955), German theoretical physicist, developed general relativity, the modern theory of gravity, and made massive contributions in other fields of physics.

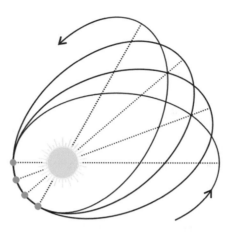

Fig. 2.1 The change of the closest point of Mercury's orbit around the Sun, the *perihelion*, over several orbits. *Image: public domain*

seconds and the flattening of the Sun causes a shift of 0.02 arc seconds and this leaves a 43 arc seconds per century error to be explained.[5] In 1915 Einstein found that his theory of gravity predicted the correct result to within the estimate of measurement errors.

At this point a naive approach to the way physics works would have predicted that Newton's theory would be dropped in favour of Einstein's. A number of factors influenced the next developments. In particular Germany was at war and so the distribution of scientific papers was slow or non existent and physicists, quite reasonably, expected Einstein's calculations and the observations to be repeated a number of times by independent physicists before being adopted. In fact, although physicists now accept Einstein's theory and the consequent perihelion shift correction Newton's theory of gravity is often used because it gives a good approximation for many calculations.

Contrary to the Popperian viewpoint that would discard a theory that has been falsified, in this case by the wrong result for the Mercury perihelion shift, Newtonian gravity theory was not immediately rejected, instead the two theories now run in parallel since Einstein's gravity theory is much more complicated to use and unnecessarily accurate for use on smaller scales and for lower velocities. In these simpler situations Newton's theory provides a good approximation. However, as we have seen above there are many situations where Newton's theory gives the wrong answers.

---

[5]The arc second is an "old" unit to measure fractions of a degree. One degree is split into 60 arc minutes, and an arc minute is then divided further into 60 arc seconds, hence we have 3600 arc seconds in $1°$.

We should highlight here that all scientific theories are provisional, but not arbitrary. Hence, Einstein's theory has replaced Newton's, but Newton's theory is "contained in Einstein's theory", and as pointed out, is still perfectly good to use on smallish scales or when gravity is weak.

2.4   Changing Theory in Physics

The example above of the perihelion shift illustrates an important feature of the scientific theory building as described by T. S. Kuhn[6] in his book "The Structure of Scientific Revolutions" published in 1962. He identified what he called a paradigm—a model or theory and all the interpretation and investment that goes with it. So above we have the Newton and Einstein paradigms for gravity. The paradigms affect how the subject is approached and even the questions that are asked and how work is done and the conceptual backing it has. According to Kuhn physics advances through paradigm shifts which are usually resisted initially. This is at variance with the Popper model that once an experimental measurement fails to conform to the theory the theory has to be rejected. For many years Einstein's theory was resisted but the paradigm shifted and although Newtonian gravity theory is still used since it is simpler and a good approximation for small masses and low velocities compared to the velocity of light.

The study of cosmology, itself, has undergone, in some sense, a major paradigm shift since the 1970s. Prior to that the subject was based on meagre observations and was largely mathematical and theoretical. Presently there is almost a surfeit of data from surveys and other observations coming from major improvements in technology. Cosmologists working today have in addition to the mathematical and theoretical technology to deal with masses of data and instead of 'just doing sums' they need the latest powerful computers and up to date data analysis methods. Looked at from the viewpoint of the cosmologist the job has changed completely as a result of a shift in the paradigm for doing cosmology. We will come back to the technology and the way science has developed in more detail in the chapters which follow.

It is a key part of the scientific method that studying the theory suggests new usages and leads to new discoveries, experiments and tests. The theory has the further useful property in that it contributes strongly to the interpretation of

---

[6]Thomas Samuel Kuhn (1922–1996), American physicist and philosopher of science.

new experimental discoveries. Also data suggests new ideas, and observations influence the questions we ask.

An essential problem in cosmology is that we cannot carry out actual experiments to test our models of the universe directly. Unfortunately, we cannot build a universe sized laboratory. Let us study as an example the universe on the largest scales and the formation of structure, on these scales, that is the distribution of galaxies and clusters of galaxies. What we can do is analyse the observational data collected by astronomers and look for patterns in the way the galaxy clusters and galaxies are distributed. Then we can deduce from this data the kind of models that lead to these particular patterns in the distribution of galaxies. But we cannot test alternative scenarios as required by the scientific method since we only have one set of data. On the one hand, fortunately, this gives us a good working model, which agrees *by design* with all of the data. But on the other hand, it is then not possible to test the model independently with different data, as we will have "used up" all the data in designing the model. To invoke the scientific method we need to do experiments, for example change the setup of our experiment and use it to test the effect on our models. For example, how does the distribution of galaxies change if gravitation is not given by Einstein's theory of general relativity but something else.

One way to overcome the problem of having just one data-set, or one universe, to work with, is to split the data into different data sets. For example use half the data to come up with a model, and use the other half to test it. This is actually a viable option, but has its limitations, as we have assumed that the data in both halves is equivalent, which might not necessarily be the case, it is an assumption.

Going back to our example of structure formation and the distribution of galaxies, we can use observations of different regions of the sky to derive our model and then to test it. But this assumes that the universe is similar in different regions. Cosmologists make indeed this assumption, and assume that the universe is the same everywhere on very large scales and refer to this as the "cosmological principle". But to test this principle is rather difficult as we saw above. We will return to the cosmological principle in later sections (see Sect. 6.4.2.4), and what observations have led to its acceptance.

Also, often there is simply not enough data available to split it up. At least this was the case in the past.

To overcome most of these problems astronomers have developed software, computer codes, to do large scale simulations of the evolution of parts of the universe using the data available. These not only enable the model to be tested but they can be used to experiment with the values of particular parameters in the model.

2.5   Cosmological Simulations

Of increasing importance in modern physics and also in cosmology are simulations. We will only discuss the basic ideas behind cosmological simulations here, and will come back to some of the underlying physical ideas in greater detail later on.

Simulations can be thought of as "experiments in a computer". Since we are working in this case in a virtual world which is in a machine, we have far more freedom to design our experiment, than if we had to conduct an experiment in the real world. We can therefore pick a region of interest without the usual inhibitions such as lab-space, time and personnel available, and of course the costs involved. Unlike a real experiment, we can also choose the governing physical laws for this region. Next we have to decide what initial conditions to choose, that is the configuration of the system at the beginning of the simulation. Then our powerful computer, controlled by its software, will evolve the system forwards in time from the initial conditions to some later time that we also specify. We can study this evolution and the final state of the system we are simulating, usually with the help of other specialised software.

Before we discuss computer simulations in cosmology, we would like to illustrate the difficulties in simulating complex physical systems using an example from an area outside of astrophysics—but closer to home, namely weather forecasting and on larger scales, climate science. Both are closely related and require the modelling of the processes in Earth's atmosphere and on the ground, but differ considerably on the time and length scales involved. It is fairly safe to assume that in these examples the underlying physics governing the system, Earth's atmosphere, is understood, much better than the physics that governs the evolution of the universe.[7] Nevertheless we face similar issues and problems in these examples as in cosmological simulations: what initial conditions do we choose, what is the level of approximation we can get away with, that is, what do we have to include in the simulation and what can be left out. The simpler our model, the easier it will be to simulate it on the

---

[7]We will however see in Chap. 6, that the forces at play over large periods in the history of the universe, are the same to the ones governing the weather, with some simple extra ingredients.

computer. The fewer approximations we make, the more complicated and computationally intensive (or expensive) the simulation will become.

If we are trying to forecast the weather, we are usually interested in spatial scales of tens or hundreds of kilometres, and time scales of days or weeks. As initial conditions we can pick the weather a couple of days ago, or even now. Climate is the weather on global scales, hence to model the climate we have to consider much larger spatial scales, ideally the planet's total surface, including all of the atmosphere and also the oceans, and work on time scales of several hundred thousand years. The initial conditions are therefore set some considerable time in the past. In both cases however, the governing equations should describe the weather phenomena, that is describe the behaviour of Earth's atmosphere and its interactions with Earth's surface, and subject to energy received by the Sun etc. From the successes and the controversies surrounding both of these areas of research, despite its "closeness" (it is literally in front of our door), we can appreciate how much more complicated the simulation of the whole universe and its contents must be.

Let us now return to simulations in cosmology. Here we will have the same problems as in other branches of physics, as pointed out above, when we set up our "experiment" in the computer. However, here we face several problems unique to cosmology. The region we would like to pick for the simulation is easily found: all of space. This is of course very difficult or impossible, since the universe is very large or even infinite. Hence we have to "make do" with a rather large region of space. The starting point of our calculation is likely to be the earliest point in time we trust our theories to work. We could for example pick a time very early on, during inflation, a fraction of a second after the beginning. Or we could pick a much later time, after the universe has cooled sufficiently for matter and radiation to have decoupled roughly 380,000 years after the beginning. This is the choice many Large Scale Structure simulations make, as this allows them to focus on the matter, and neglect the by then relatively unimportant radiation content. Choosing suitable initial conditions then turns out to be another problem, as we have to specify the positions and velocities of all the constituents of the simulations. This information will be the outcome of a separate, rather involved, calculation itself.

Finally what approximation do we allow for when we specify the physical laws that govern our universe in the computer. Ideally we would like to implement all physical laws into a massive computer code, then watch the simulation unfold. This way we can be sure we have not omitted some crucial detail. This is not possible, unfortunately, since, the time and length scales in cosmology are simply too large to model the universe as a whole in any detail. Using this brute force approach, modelling simply everything, would already be prohibitively expensive—in terms of computer processing time and therefore money—even if we try to model only a spoonful of matter.

Another problem unique to cosmology is, that we might not know all the relevant physical laws! We are fairly confident that in familiar, "every day" settings we know the relevant laws, but what about the beginning of the universe, when conditions were very different to today? Again, we have to make some assumptions and then decide from the results that we picked the correct laws. Here the difficulty is to decide whether the results we get are due to the laws or are due to the approximations we have made, and might be very different if we had made a different choice. The results will depend on the approximations and the model, in addition to all the numerical considerations.

The complexity of realistic physical systems therefore requires us to make simplifying assumptions. This is not restricted to cosmology, it is the basis of most computer simulations (and indeed at the basis of most calculations in physics). For example, in many or most settings in fluid dynamics it is not necessary to model each particle or atom. Instead we can treat a vast number of particles like a single virtual object, and follow for example the motion of a cubic metre of gas instead of single gas atoms. This will lead to a huge simplification, considering that a cubic metre of air contains roughly 10^{27} molecules (at standard conditions).

Nevertheless, even "simple" systems are complicated to simulate and the software and hardware, the numerical codes and the computers, have only recently become powerful enough to model the climate of our planet on large scales, both in space and time, or the weather, on smaller scales, with some accuracy. In current weather simulations a resolution of 1 m would still be far too fine grained, and researchers therefore use resolutions of the order of kilometres.

Similar problems face researchers in cosmology, even more so, as we have to allow for the larger time and length scales relevant here. For numerical simulations in cosmology the resolution has to be suitably adapted, with the "smallest" length scales measuring millions of parsecs. What do the results of these simulations look like? Figure 2.2 shows three snapshots of a recent simulation from the Millennium simulation, taken at 650 million years, 2.9 billion years, and 13.8 billion years after the beginning.[8] The size of the side

---

[8]To simulate the evolution of a huge volume of the universe is only possibly by allowing the computer to evolve the governing equations using time-steps of the size of years. This makes it possible to simulate billions of years in the history of the universe in a couple of weeks of computer time. That the results nevertheless agree with the real universe, can be checked by using simple test cases against which the simulated results are compared.

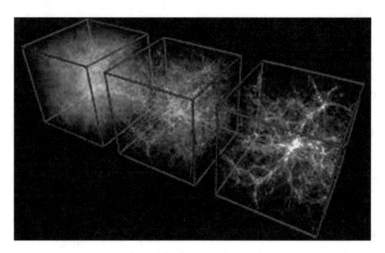

Fig. 2.2 Three "snapshots" from a simulation studying the formation of large scale structure in the universe. The snapshots are taken at 650 million years (left box), 2.9 billion years (middle), and 13.8 billion years (right box) after the beginning. The size of the side of the cubes is roughly 430 million lightyears, denser regions are more opaque. We see that small density inhomogeneities, that is small differences in the distribution of the density of the matter, already present at early times, get amplified through the gravitational instability, leading to the formation of structure, the "cosmic web". Note, that the size of the box is growing with the expansion of the universe. *Image credit: Volker Springel*

of the cubes is roughly 150 million parsecs. The whiter or cloudier regions are denser, and we start with small over-densities initially. Then, due to gravity, the over-dense regions attract more matter and become even denser. Finally, a distribution of galaxies similar to what we observe in galaxy surveys today emerges. The length scales in the simulation, including the size of the box, are chosen to match the expansion of the universe (we will discuss these "comoving" with the expansion coordinates later on in Sect. 7.3). Therefore we do not see the expansion of the universe, although the box size increases by a factor of 6 from the beginning to the end of the simulation.

We can compare the results of the simulation to the observed distribution of galaxies, see Fig. 1.4 in the previous chapter. We can see, that both in the real and in the simulated universe, the galaxies on large scales are not randomly distributed, but aligned along filaments, which cosmologists refer to as the "cosmic web" (there are statistical tools that allow comparisons to be made which are more rigorous than simply confirming by eye that the "look" of the galaxy distribution is the same in both cases). On scales much larger than the box size of the simulation, the distribution of galaxies appears however smooth (we will return to this in later chapters).

We should stress here again that, to an extent, we can only get out of a simulation what we put in: we choose the physical laws at play and assume we know how to pick the right initial conditions. Only after making these choices and successfully running the simulation, in itself a highly non-trivial undertaking, the output of the simulation can be compared with the observations.[9] Again, we can only check whether they disagree or that both are consistent, we cannot "prove" that our assumptions are correct.

We have seen how physical laws are developed through the interplay of theory and experiments, and in particular in cosmology, by confronting theory with observations. Before we discuss what physical theories we need for cosmology, we will have a look at how the relevant observations are made and what kind of observational data we have at our disposal, in the next chapters.

---

[9]We can of course also compare the numerical results with other theoretical calculations, arrived at using pencil and paper. This will allow us to check the numerical procedures—we haven't discussed the numerical and computational errors—and also check our pencil calculations and the approximations we made in them.

3

What Observations Do We Use?

In this chapter we will discuss what kind of observations informed our picture of the universe. To this end we will discuss briefly electromagnetic radiation, the various forms of "light", and also briefly touch on "non-standard messengers" such as neutrinos and gravitational waves. We can then ask, how are these observations actually made? This will allow us to highlight some recent telescopes and experiments.

3.1   Introduction

Astronomy and cosmology are enjoying a golden period of discovery and so there are currently numerous observational projects in operation and exciting new ones about to come on line in the next couple of years. In view of the fast changing nature of astronomy we will not focus on particular observational programmes and equipment, instead we will try to pick typical or, in our opinion, particularly important projects or equipment to discuss and illustrate the processes. Perhaps we should add that the technology and methods to undertake observations will be extended and improved in the future, but the key role of observations in astronomy will remain the same.

© Springer Nature Switzerland AG 2019
K. A. Malik, D. R. Matravers, *How Cosmologists Explain
the Universe to Friends and Family*, Astronomers' Universe,
https://doi.org/10.1007/978-3-030-32734-7_3

3.2   What Reaches Us

Let us start by asking what do we actually mean by "making observations"? Here we mean the gathering of information, and since we are interested in the universe, we are in particular interested in information that reaches us from far away. The information can reach us in various forms, the most familiar of which is probably electromagnetic radiation, such as light, but also radio waves or gamma rays. Non-standard messengers, such as neutrinos and gravitational waves, are also beginning to play an important role in astronomy.

In the following sections we describe what kind of data, relevant to cosmology, is collected by astronomers and what observational tools they use. The interpretation and significance of the observations is then the subject of much of the rest of this book. We use the terms "telescope" and "observations" in a very broad sense in this chapter and elsewhere in the book. When we refer to a "telescope" we mean an instrument that enables astronomers to observe, in a very general sense, objects in the sky and to collect data (observations) from them.

3.2.1   Different Forms of Light: Electromagnetic Radiation

As mentioned above, electromagnetic radiation is at the moment our most important source of information about the universe. So what is this radiation?

Electromagnetic radiation, or electromagnetic waves, are electric and magnetic fields changing periodically and propagating across space and time, see Fig. 3.1. We will discuss these fields in more detail in Sect. 6.2, and here simply highlight some relevant details necessary to make observations.

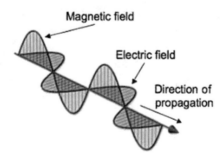

Fig. 3.1 An electromagnetic wave, the electric and magnetic fields that make up the wave oscillate at right angles to the direction in which the wave travels. The fields themselves oscillate in planes at right angles to each other. The distance from the maximum of one wave to the maximum of the next is the wavelength of the wave. *Image credit: Juming Tang*

The electromagnetic waves travel at the speed of light, and each wave has a wavelength and a frequency (or period). The speed of light is constant and very large, 300,000 km/s, but finite. The frequency of the waves and their wavelength are proportional, that is long wavelength radiation has low frequencies, short wave lengths high frequencies. All wavelengths of the electromagnetic radiation are known as the electromagnetic spectrum: from the longest to the shortest wavelengths the radiation is divided into classes based on the wavelength of the electromagnetic waves, see Fig. 3.2.

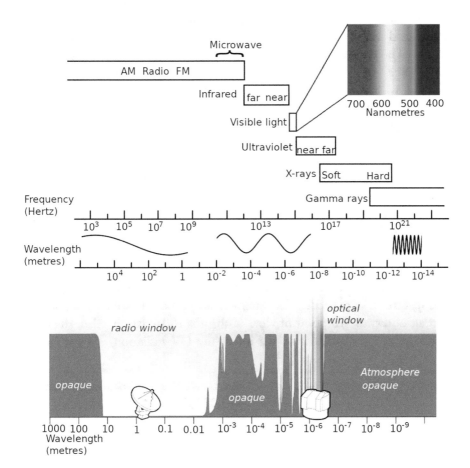

Fig. 3.2 The electromagnetic spectrum: at the top of the figure are the different frequency bands of electromagnetic radiation, starting at low frequencies with the radio band, then with increasing frequency the infrared, visible and ultraviolet bands, and finally at very high frequencies the X-ray and gamma-ray bands. Next in the figure we have the frequencies and the wavelengths corresponding to these bands. Also shown at the bottom are the regions of the spectrum in which the atmosphere is transparent for the radiation. *Image Credit*: Malik and Matravers; atmospheric opacity image ESA/Hubble (F. Granato)

The electromagnetic fields that make up the radiation oscillate in planes at right angles to the direction the wave travels. The planes in which the magnetic and electric fields oscillate are also at right angles, see Fig. 3.1. Radiation is made up of lots of individual waves, with the fields oscillating in random directions. If the fields oscillate mainly in a preferred plane, the electromagnetic radiation is "polarised".

At the long wavelength end of the spectrum we have radio waves, familiar from our AM or FM radios, with wavelengths ranging from roughly 100 km to 1 m. Next microwaves, as for example used in the familiar microwave ovens, have wavelengths that range from 1 m to 1 mm. Infrared radiation, usually experienced as "heat", comes next with wavelengths from roughly 1 mm to 780 nm. The part of the spectrum we are probably most familiar with, visible light, ranges from 780 to 380 nm. After visible light we have ultraviolet radiation with wavelengths from roughly 380 to 100 nm. Subsequently we have X-rays, with shorter wavelengths ranging from 10 to 0.001 nm, and finally gamma rays at even shorter wavelengths than 0.001 nm (see also Table 3.1).

Astronomical objects between them emit electromagnetic radiation across all wavelengths in the spectrum with various intensities, hence ideally, we would like to make observations across all wavelengths. But not all regions of the spectrum are equally accessible to us. Radiation passing from some astronomical object to us will have to pass through the Earth's atmosphere. If the light from for example a star passes through some turbulent patches, this will lead to changes in light intensity. This causes the familiar "twinkling" of stars.

Even more important for us, the atmosphere blocks a lot of the electromagnetic spectrum. In particular the water vapour in the atmosphere is an obstacle from an observational point of view, it absorbs IR radiation. Also, the ozone in the upper atmosphere blocks a lot of the UV radiation. This is bad for

Table 3.1 The wavelength ranges of different parts of the electromagnetic spectrum

Radio	100 km–1 m
Microwave	1 m–1 mm
Infrared	1 mm–780 nm
Optical	780–380 nm
UV	380–100 nm
X-ray	10–0.001 nm
Gamma-ray	Less than 0.001 nm

A nanometre, abbreviated "nm" is a billionth, a millimetre, abbreviated "mm" is a thousandth of a metre

astronomers, but good for life on Earth, since it means that we are protected from most of the high energy radiation.

There are however two convenient observational windows in the radio and in the optical region of the spectrum, if we want to do observations from some non-exotic place on Earth. We can improve on these windows by going "high and dry", where the atmosphere is thinner and there is less water vapour in the atmosphere. This lead astronomers to build observatories high up on mountain tops, the Atacama desert, and the South pole, where the air is thin to minimise the distortions, and dry to minimise the absorption of radiation through water vapour. These atmospheric conditions also allow observations in the microwave part of the spectrum.

Although the speed of light is large it is finite so we only see an object at a time in the past, for instance we only see an event on the Sun 8 min after it happened because light takes 8 min to get from the Sun to us. The nearest star to the Sun is Proxima Centauri which is about 4.2 light years away, in other words light takes 4.2 years to get from Proxima Centauri to us. Clearly looking out into space is accompanied by looking back in time. In the light of this it is not all that surprising that astronomers can talk of "seeing" the Big Bang—they mean that they have detected radiation that has come to us from the time of the Big Bang.[1] Since we know the speed of light we can use the time or distance to measure how distant a star or other celestial object is from us. For instance the distance to Proxima Centauri is 4.2 lightyears or if we prefer it, 3.9×10^{13} km, since the speed of light is 9.46×10^{12} km/year, or 9460 billion kilometres per year.

Later on in Chap. 4 we will discuss distances and distance units encountered and used in astronomy and in cosmology in some more detail.

3.2.2 Non-standard Messengers: Neutrinos and Gravitational Waves

Although most observations to date rely on the gathering of electromagnetic radiation, astronomers are beginning to use also "non-standard messengers". Let us first define what we mean by "messenger" in this context. This can in principle be any physical entity that carries information from an emitter or source and reaches a receiver or observer. The most common example is an electromagnetic wave, what we might therefore refer to as our "standard

---

[1]The term "Big Bang" is not very well defined. Here we mean very early on in the history of the universe. Throughout the book we use the phrase "the beginning" to denote the beginning of the universe.

messenger". But we are not restricted to only consider electromagnetic waves, since the actual physical nature of the "messenger" carrying information to us is not that important, as long as the information can reach us over the vast, astronomical distances we are interested in. A good or useful messenger for astronomers is therefore any carrier of information that is not diverted by the forces acting on it and diverting it on its journey, and it should also not interact with or be absorbed by the stuff filling the space between the emitting source and us, the receiver. Once it has reached us, it should however "hand over" the information, so at the end of its journey it should interact with our observational equipment. Finally, a good messenger should travel as fast as possible.

In the case of electromagnetic radiation or photons, which are our standard messengers, these conditions are partially satisfied: electromagnetic waves are only affected by gravity on their journey from, for example, distant galaxies to us, but not by the other forces of nature[2] (see Chap. 6 for a discussion of these forces). Since electromagnetic waves travel with the speed of light, the information will reach us as quickly as possible. However, photons interact with matter and get absorbed for example by interstellar gas or Earth's atmosphere.

The conditions mentioned above limit possible candidates considerably. Although any particle or wave could be used as a messenger, ideally the messenger, should be affected by as few forces as possible, and interact as little as possible on its journey to the observer. It should also carry little mass, so that it can travel at a speed close to the speed of light (photons are massless).

However, one of the qualities that makes a good messenger, namely that it travels without too much interaction from the source to us, also poses a problem: how can we observe it, if it doesn't interact or cannot be absorbed by a detector? Unfortunately, if the messenger doesn't interact at all, we can't use it. So it should at least interact weakly, allowing us to actually detect it.

Two at present popular non-standard messengers that satisfy these conditions are neutrinos and gravitational waves. We will describe their use in observational astronomy below.

---

[2]Electromagnetic waves or photons can also be affected by other photons, but this effect is very small. Overall the universe is "charge-neutral" and there are therefore no electric forces on large scales, and the magnetic fields on large scales are extremely tiny and have a very small effect on the photons.

3.2.2.1   Neutrinos

Neutrinos are probably the most prominent example for particles other than photons that reach Earth from astronomically large distances. Neutrinos are nearly massless, weakly interacting particles. Due to their small (close to zero) mass, they travel with nearly the speed of light. Neutrinos do not carry electric charge, and therefore do not interact electromagnetically (they are not affected by electric and magnetic fields).

Besides being affected by gravity, like photons, they only interact weakly with matter on their way to Earth. To get an idea of what weakly interacting with matter actually means, we note that a neutrino can pass through several lightyears of lead before it interacts with a lead atom and gets absorbed.

However, they can be used for observational purposes despite their being weakly interacting, because there so many of them. For example, the fusion processes that power stars produce copious amounts of neutrinos. Due solely to the Sun's fusion processes, every second roughly 65 billion neutrinos pass through an area of one square centimetre (on Earth). However, because they only interact weakly, these neutrinos have nearly no observable effect. We need to build truly gigantic detectors and give the neutrinos many opportunities to interact, in order to observe and measure just some of them.

Because neutrinos interact weakly, they can bring us information from regions that are opaque to electromagnetic radiation. We can for example get information about the inner region of stars, where the fusion processes take place, by observing the neutrinos that reach us from there.[3]

But there are also sources other than the normal fusion processes of stars that produce a neutrino signal. During a supernova explosion large numbers of neutrinos are produced. These neutrinos have already been observed by current neutrino detectors.

Another source of neutrinos, directly relevant for cosmology, is the Cosmic Neutrino Background. We already mentioned the Cosmic Microwave Background, by which we mean the photons that reach us from a time when the universe was roughly 380,000 years old. The Cosmic Neutrino Background formed much earlier, when the universe was just 1 s old. Before this time the universe was so dense and hot, that the neutrinos interacted with other particles present in the primordial soup at this time. As the universe expanded it also cooled down, and was after roughly 1 s cool enough for the neutrinos to

---

[3]Photons are also produced in these processes, but they interact, get scattered, many times before they reach the surface of the star. We can therefore only "see" the outer layers of the Sun, for example.

stop interacting with the other particles. If we could observe these neutrinos, we would be able to get a snapshot of the universe when it was just 1 s old, just like the Cosmic Microwave Background presents us with a snapshot of the universe 380,000 years later. Unfortunately these neutrinos are even more difficult to measure than the photons of the Cosmic Microwave Background.

In principle we can also use other particles as messengers, but they should conform to our "wish list" above, that is be light otherwise they clump, if they are really heavy, and are slow, and they shouldn't interact with "stuff" on their way to us.

3.2.2.2   Gravitational Waves

Let us finally consider gravitational waves, the latest addition to our group of messengers used in astronomy. Gravitational waves are wiggles in spacetime. They are generated in the most violent events in the universe, such as during inflation, supernova explosions and the merging of black holes.

Like electromagnetic waves, gravitational waves travel with the speed of light and have a period of oscillation (or frequency) and a wavelength, and like electromagnetic waves they are transverse that is the oscillation is perpendicular to the direction of motion, another example being water waves on the surface of a pond. However, in the case of gravitational waves it is not the amplitude of the electric and the magnetic field that oscillates periodically, it is the distortion of spacetime. These distortions lead to a periodic change of the distance between objects as the gravitational waves passes these objects. These changes in distance can be measured and allow astronomers to detect the waves when they pass through their detectors.

Because gravitational waves are distortions of spacetime, they do not interact with matter and can therefore travel unhindered across vast distances. This also means that they can carry information from regions no other messengers can reach us from, for example allowing us to get information directly from the epoch of inflation. The gravitational waves generated during this epoch will have travelled to us from these incredibly early times.

The wavelengths of gravitational waves are roughly comparable to the size of the emitting region or object, very different to electromagnetic waves, where in astronomy the wavelength is usually much smaller than the object emitting the radiation.

Gravitational waves are very weak, at least by the time they reach observers on Earth. The typical change in distance between two masses due to the gravitational wave is as small as 10^{-22}, which means for example that the

LIGO observatory with a distance between its mirrors of 4 km, has to measure a change in length of roughly 4×10^{-19} m. When we compare this to the diameter of a typical atom of 10^{-10} m, we get an idea of the daunting task of making this measurement! We will describe the detectors in more detail below in Sect. 3.3.4.2.

Gravitational waves were observed directly for the first time on 14 September 2015, when the LIGO observatory observed the merger of two black holes. For more on gravitational waves and gravity we refer to Sect. 6.4.2.3.

3.3 How Observations Are Made

In this section we will introduce and describe some of the tools available to astronomers to make detailed observations and collect data. We will follow the common practice and call all detectors or receivers telescopes even if they do not operate in the visible electromagnetic range and do not even have lenses or mirrors. The telescopes are, in practice, distinguished by the wavelength range in which observations are made or data is collected, whether or not they use other "messengers" than electromagnetic radiation, and whether they are on the ground, attached to a balloon or in space. To give some structure to our descriptions we will follow a mini-historical view, but follow history not too closely.

Until the invention of telescopes in the seventeenth century observations were made by eye. This restriction meant that the models and descriptions that could be developed were limited. Nevertheless it is amazing how much was learnt about the solar system and the locations of bright objects, mainly stars within our galaxy, by those early astronomers. For example the planets and our constellations of stars were identified and named by the Babylonian about 3000 years ago. A proper understanding of the motion of the planets was delayed by the conviction that the Earth was at the centre of the "universe". Even so Ptolemy,[4] who lived in the second century in Alexandria in Egypt, was able to predict the positions of the planets with his Earth centred model which used "epicycles".[5]

---

[4]Claudius Ptolemaeus (about 100–170), Roman Astronomer and Mathematician, influential works on geometry and astronomy.

[5]In this model, the motion of the planets is explained by the planet moving on a small circle, the centre of which moves on a larger circle. The "epicycle" is this movement of a circle on a circle.

3.3.1 Optical Telescopes

The invention and use of early, optical telescopes by Galileo,[6] among many others, at the beginning of the seventeenth century changed astronomy completely. Using convex lenses and concave mirrors observers were able to determine details of distant luminous objects, for example the planets, and show that they orbit around the Sun not the Earth.

In the following years the telescopes were refined and their sizes increased to capture more light from the object being studied. For instance William Herschel[7] and his sister Caroline Herschel[8] constructed a telescope over 12 m long in their garden in Slough near London in 1789. It was used for 50 years, though only a model exists today. Using the telescope they identified smudges of light, which they called nebula. These nebula were later recognised as clusters of stars, and some as galaxies.

Initially most astronomers used telescopes with lenses, usually referred to as "refractors". Since it is easier to build large light focusing mirrors than lenses, these days nearly all telescopes use mirrors as their primary light collecting device. A lens can only be supported at its rim, and also has to be large and massive and is therefore very heavy. A curved mirror on the other hand can be supported on its entire back side, and the light gets reflected from its surface, which therefore only needs a thin, reflective coating, see Fig. 3.3. Also lenses suffer from "chromatic aberration" (different colours get refracted differently), whereas mirrors do not. The telescopes astronomers used were eventually redesigned so that an observer could attach various cameras and electronic measuring devices according to what was being studied, instead of having to rely on the naked eye.

Nowadays the old image of the astronomer making observations sitting on a stool at the top of a ladder in a telescope dome has given way to an astronomer sitting at a computer, downloading data from a computer attached to the telescope perhaps many miles away. The astronomer looking at the image is replaced by an electronic device attached to the eyepiece of the telescope which scans the image and produces an electronic version which can be used to extract those features of the data that are wanted. In many ways the electronics

---

[6] Galileo Galilei (1564–1642), Italian astronomer, physicist and engineer, groundbreaking works in physics and astronomy.

[7] Friedrich Wilhelm Herschel (1738–1822), German astronomer working in London, made early maps of the universe.

[8] Caroline Lucretia Herschel (1750–1848), German astronomer working in London, made early maps of the universe.

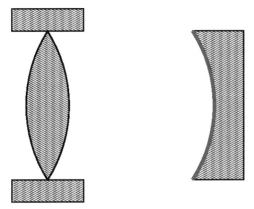

Fig. 3.3 A lens can only be supported at its rim, a mirror (here red) can be supported on its entire back side

are similar to a superior digital camera. Because it is no longer necessary for an astronomer to be present, observations over long periods of time are now possible. Also the telescope can be set to automatically track objects as the Earth moves and to download information at set times. Taking this idea one step forward, recent surveys, for example the Sloan Digital Sky Survey (SDSS) and 2dF, the "Two-degree-Field Galaxy Redshift Survey" (which we discuss in more detail in Chap. 4), have used relatively small telescopes to comprehensively survey the sky. Night after night the telescope is used to track across a fresh part of the sky by using the Earth's rotation. In this way millions of galaxies and stars have been observed, their positions determined, and other data collected from them. More details are given in the next chapter where we discuss what is collected and how it is analysed.

Over the years there has been a tendency to build ever larger telescopes, culminating in today's giant telescopes. This is because the larger the collecting area of the telescope, the fainter the source of the radiation that can be studied. Simply put, the larger the collecting area the more radiation is collected. However there is a major problem with building telescopes with larger lenses and mirrors to get greater resolution which is that the weights of large lenses and mirrors lead to distortions in either or both. The effects of these distortions on the image have to be removed. One way to overcome this problem is to use mechanical actuators to reshape the mirror to compensate for the distortions which it is difficult to achieve efficiently.[9]

---

[9] Some theoreticians—not us—joke that the mirrors are reshaped to produce the images observers want.

Instead of having one giant telescope we could build several smaller telescopes, either a few as in the case described below, or of several hundreds spread out over a large area, as for example in the case of the SKA ("Square Kilometre Array", see Sect. 3.4.1). The output of the single telescopes can then be combined through optical fibres or electronically, if the telescopes are focused on the same object. The resolution and brightness of the image are the same as if there was a single telescope with a mirror the size of the telescope array that is the area covered by the telescopes. But the telescopes can also be used independently, allowing the observation of different objects at lower resolution. More details of how this is achieved are given in Appendix A.4. An example of an optical telescope that uses an array of receivers is The Very Large Telescope (VLT) which is located at the European Southern Observatory (ESO) at the Paranal Observatory in Chile, Fig. 3.4. The VLT consists of four individual telescopes, each having a primary mirror with 8.2 m diameter. It operates in the optical and near infrared wavelengths. The images taken by the VLT are truly amazing, see Figs. 4.3 and 4.4 in the next chapter.

Fig. 3.4 The Very Large Telescope (VLT). Only three of the four Unit Telescopes of the VLT are shown here, getting ready on top of Cerro Paranal, in Chile. *Image: ESO*

A valuable extension to ground based observations is to place a telescope on a satellite, such as the Hubble Space Telescope (HST). Placing the telescopes outside of Earth's atmosphere has the advantage of avoiding the distortions through atmospheric turbulence, and absorption by dust and water vapour (at some wavelengths). As an example we mention here the Hubble Space Telescope (HST), which was launched in April 1990 and is still taking data. Its primary mirror has a diameter of 2.4 m, and it observes visible, ultraviolet, and near-infrared wavelengths. It is the most famous space based telescope which has been serviced in space four times. Some remarkable images have been captured by the HST so we have included two examples in the next chapter, Figs. 4.2 and 4.6.

The successor to the HST, the James Webb Space Telescope (JWST), is currently being built and scheduled to be launched in 2021. It will have a mirror of 6.5 m in diameter, and will take images in the visible to mid-infrared wavelength range.

In retrospect one feels that it should not be a surprise that many astrophysical objects, for example the Sun, or active stars emit radiation in more wavelengths than those in the visible part of the electromagnetic spectrum. In fact they do which provides additional opportunities for making observations of astrophysical phenomena. We receive some infrared radiation (heat) fairly regularly, we can feel it on our skin, but it is hard to extract useful information from it here on Earth, because of absorption and distortion due to the atmosphere. Looking at the spectrum of electromagnetic radiation in Fig. 3.2 (and Table 3.1) it seems natural that the next wavelength (frequency) we should look at is the "near infrared",[10] that is, radiation with wavelengths of about 10^{-7} m. A problem with using infrared radiation for observations on the ground is that it is strongly absorbed in the atmosphere, especially by the moisture in the air. Also additional infrared radiation is produced by the scattering of the heat from the Sun in the atmosphere. For these reasons there are few ground based infrared telescopes and those there are, for example the VLT, work at the near infrared.

---

[10] "Near infrared" refers to radiation with wavelengths near to the optical range.

3.3.2 Radio Telescopes

It was not until the 1930s that radiation in other wavelengths was noticed with applications in astronomy in mind. In 1932 Karl Jansky,[11] who worked for the Bell Telephone Company, reported on an investigation he had carried out to find the sources of static affecting long distance radiotelephone communications. He identified one of the sources of the radio interference as a celestial object. Unfortunately Jansky's work was largely overlooked originally, but Grote Reber,[12] heard about Jansky's work and in 1937 built his own radio telescope in his back yard. After the second world war had ended, significant further work on radio telescopes was done. This was an ideal time for radio astronomy to begin in earnest, because redundant radio and radar equipment and people with the skills to use it were available after the end of the second world war. As radio telescopes are less familiar than optical telescopes we'll briefly describe them. Basically the early instruments were radio wave receivers or antenna tuned to particular wavelengths.

As with optical telescopes, the largest receivers possible are needed for the radio telescopes because radio signals are even weaker than optical ones. Following the optical telescopes, two main designs are used. The first one has a large single dish, which reflects the incoming radio waves to an antenna at the centre of the dish. The dish has a specific shape (it has a parabolic cross section) that allows it to collect the radio waves over its whole area and to reflect them to the central antenna or receiver for further processing. The large radio telescope at Jodrell Bank is of this type, see Fig. 3.5. The whole telescope dish can be steered to observe different objects.

The other type, which is now becoming more popular, has an array of small receivers distributed over a large area, sometimes a very large area. Modern arrays may even range over a number of countries, for example LOFAR, or the Low-Frequency Array, a telescope based in Holland, where all the signal processing is done, has receivers located in Holland, Germany, the UK and there are plans to extend it further. Some of the LOFAR local receivers are similar to a very large grid of radio aerials (roughly 20,000 simple dipole antennas). In other places they each take the form of a small version of the Jodrell Bank structure with small dish. The signals from all antennas are transmitted to a central computer facility, where the signals are

---

[11]Karl Guthe Jansky (1905–1950), American physicist and engineer, pioneering work in the area of radio astronomy.

[12]Grote Reber (1911–2002), American engineer, pioneering work in the area of radio astronomy.

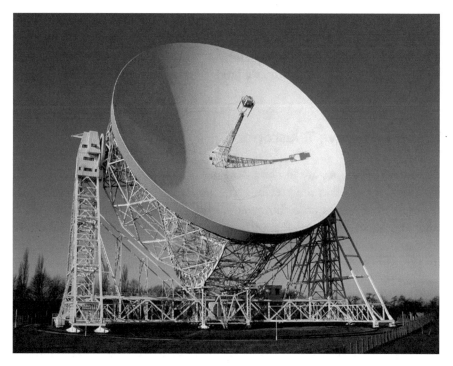

Fig. 3.5 A typical radio telescope: the Lovell telescope at Jodrell Bank. The radio dish has a diameter of 76.2 m. *Image credit: Jodrell Bank Centre for Astrophysics, University of Manchester*

combined, electronically turning the individual antennas into a single, very large radio telescope. LOFAR's receivers work at radio-frequencies between 10 and 240 MHz (although probably familiar from old fashioned radio dials, 1 MHz is 1 million Hertz, and 1 Hz means one oscillation per second).

The central receivers for both types of radio telescope collect the incoming radiation and convert it into an image. These images are usually displayed as contours of equal signal intensity or as false-colours on a map where each colour corresponds to the signal strength. We present as an example the image of a galaxy taken by the Very Large Array in Fig. 4.5 in the next chapter.

As an example of a working multi-antenna radio telescope we will briefly describe the Very Large Array (VLA). The Very Large Array is currently one of the largest operating radio frequency observatories in the world. It consists of 27 independent antennas, each of which has a dish diameter of 25 m. The antennas can be physically relocated to a number of prepared positions, allowing signal processing to obtain a maximum baseline of 36 km: in essence, the array acts as a single antenna with that diameter which picks up radio signals with wavelengths from 0.7 to 400 cm (or 50 GHz to 73 MHz). It is a multi-purpose instrument designed to allow investigations of a wide range of astronomical objects, including quasars, supernova remnants, black holes, and the hydrogen gas that constitutes a large portion of the Milky Way galaxy as well as distant galaxies.

3.3.3 Microwave Telescopes

Microwave telescopes are in their working principle very similar to the radio telescopes described in the previous section. Extending the range of radio telescopes to microwaves poses however various problems. As mentioned above, water vapour in the atmosphere absorbs microwave radiation. Also, there is a large amount of "stray radiation" near the surface of Earth including heat in the atmosphere, so the receiver has to be cooled to reduce its temperature, or it would be "blinded" by its own heat radiation.

Despite the problems of making observations with microwaves, Arno Penzias and Robert Wilson[13] at Bell Labs in Holmdel, New Jersey, made a remarkable discovery in 1964. They were experimenting with a extremely sensitive, 6 m horn antenna originally built to detect microwaves reflected off low level echo balloon satellites.[14] In order to measure the faint reflected waves, they had to eliminate all recognisable interference from their receiver. They removed the effects of radar and radio broadcasting, and suppressed interference from the heat in the receiver itself by cooling it with liquid helium to $-269°$ C, only $4°$ above absolute zero. When they reduced their data they found a low, steady, mysterious noise that persisted in their receiver. This residual noise was 100 times more intense than they had expected, was evenly spread over the sky, and was present day and night. They were certain that the radiation they detected which had a wavelength of 7.35 cm did not come from the Earth, the Sun, or our galaxy. In fact they had stumbled upon radiation from the Big Bang that had been predicted by George Gamow[15] and others in 1946. We will pick up this remarkable story in Sect. 8.2.3.

Observing in the microwave domain turns out to provide a window between optical and radio. In fact at metre lengths the radiation from quasars and

---

[13]Arno Allan Penzias, (1936–), and Robert Woodrow Wilson (1936–), both American physicists and astronomers, discovered the Cosmic Microwave Background.

[14]NASA launched during "Project Echo" in 1960 balloons with metallised surfaces and roughly 30 m diameter into a low Earth orbit, and reflected microwaves off the balloon back to a receiver on Earth.

[15]George Gamow (1904–1968), American physicist, influential works in cosmology.

radio galaxies dominates the picture and at infra-red wave lengths the sky background is dominated by radiation from extra-galactic sources and by faint structures in the Milky Way. Fortunately the microwave region picked by Penzias and Wilson not only lies between these two but it also lies at the peak of the wavelength of the Cosmic Microwave Background.

As pointed out earlier, the water vapour in Earth's atmosphere absorbs most microwave radiation. In order to get round this problem, astronomers have to find places where there is less atmosphere and less water vapour. Microwave telescopes are therefore best set up in high and dry places on the ground, or even higher in the upper regions of the atmosphere, or in space.

3.3.3.1 Ground Based Microwave Telescopes

Although we receive some microwave radiation even at sea level at temperate latitudes, as shown by Penzias and Wilson,[16] to reduce the absorption of the microwaves in the atmosphere we have to go to places that are high and dry, where the atmosphere is thin and there is little water vapour around. This makes the arid high plateau of the Atacama Desert and the region around the South pole ideal places, the Atacama Desert being above 5000 m altitude, and the south polar region at 2800 m being high, dry and very cold. At present ground based Cosmic Microwave Background telescopes or experiments include POLARBEAR in the Atacama Desert and BICEP at the South Pole. We will briefly describe the BICEP telescope in the following.

The BICEP microwave telescope is located at the Amundsen–Scott South Pole Station since 2005. Its main purpose is to map the temperature fluctuations in the Cosmic Microwave Background on small angular scales, but also the polarisation of the Cosmic Microwave Background. The experiment has been upgraded several times since its inception, and at present the BICEP3 telescope has 2560 individual detectors taking data at a wavelength of roughly 3 mm.

Once we move beyond the ground-level for observations, we find that a number of options and opportunities open up, including increasing the range in which the electromagnetic spectrum can be observed. Of course it all depends on price and objectives and the availability of balloons and satellites to carry the telescopes. However they open up new areas of the spectrum that are of interest to us. We will start with the simplest and cheapest options.

---

[16]Also roughly 1% of the static on old TV sets (with aerials) is due to the Cosmic Microwave Background radiation.

3.3.3.2   Balloon-Borne Microwave Telescopes

Balloon-borne experiments or telescopes are often a cost effective alternative to satellite missions. The experimental cargo, the microwave telescope in this case, is suspended below a large helium filled balloon. The balloon is then launched, usually at very high or low latitudes (that is close to the North or South poles) during the respective winter months, when it is dark and cold (and hence little heat in the atmosphere), and stays at a height of up to 40 km for up to 40 days. The payload, the telescope, is usually suspended below the balloon.

One such mission, "Boomerang", made its maiden flight in 1997 and took data in 1998 and 2003. Effectively a microwave receiver attached to a balloon which is set-up to rise to a predetermined height, 42 km, and to remain there for the duration of the observing. The experiment was set up so that as the Earth rotates the balloon circles over the Antarctic to get total (microwave) sky coverage of the Cosmic Microwave Background over a complete circuit of part of the Southern sky. This made a valuable early contribution to the mapping of the Cosmic Microwave Background (in 1998 and 2003).

A similar balloon borne telescope called MAXIMA was set up to fly in the USA in 1998 and 1999. It mapped a smaller part of the sky but at a higher resolution.

3.3.3.3   Space Based Microwave Telescopes

There have now been several satellite missions with microwave telescopes and data collecting equipment on board. The first of them was called "Cosmic Background Explorer" or COBE, launched in 1989 which determined the temperature and the black body nature of the Cosmic Microwave Background.

We will describe the second mission, called WMAP (Wilkinson Microwave Anisotropy Probe) launched in 2001 in more detail. WMAP gathered a stupendous amount of information. In fact it was ambitiously designed to determine the geometry, content, and evolution of the universe via a full sky map of the temperature anisotropies or fluctuations of the Cosmic Microwave Background. The choice of orbit, sky-scanning strategy and the instrument and spacecraft design were made to ensure a minimum of errors and to allow multi-wavelength observations and accurate calibration. The temperature fluctuation data from the WMAP observations have 45 times the sensitivity and 33 times the angular resolution of the COBE mission. This illustrates the speed of development in the observational domain in cosmology, which kept theoretical cosmologists busy, allowing them to formulate the cosmological standard model.

The most recent satellite to be launched was called "Planck" and was launched in 2009, with similar aims as WMAP but at much higher precision and also to study the polarisation of the Cosmic Microwave Background radiation. One of the highlights of the mission, an all-sky map of the Cosmic Microwave Background was released in 2013, see Fig. 1.3 in Chap. 1. The mission was a great success, the data is still being analysed, the final data release was in 2015. The data released has been extremely valuable and stealing a point from later in the book the images have proved remarkably consistent with the SDSS and 2dF ground based galaxy surveys. The satellite was decommissioned, much later than originally planned, in 2013 when it had used up all its coolant.

We will discuss the map of the temperature fluctuations made by the WMAP and Planck satellites in more detail in the following chapters.

3.3.4 Examples for Non-standard Messenger Observatories

Until recently the focus in the astronomical community was on observations using light and other electromagnetic radiation. This was mainly due to the limited technological possibilities in building detectors for other "non-standard messengers". Of particular importance here are as pointed out above gravitational waves and neutrinos. Two examples for already existing gravitational wave detectors and neutrino telescopes are Icecube and LIGO.

3.3.4.1 Icecube

We mentioned above that neutrinos interact weakly with other particles, which means that although they interact only extremely rarely, they do occasionally. From our point of view, using neutrinos for observational astronomy, of particular interest is when the neutrinos can interact with, that is bump into, water molecules in liquid or frozen form. In this case they can create charged particles, electrons and other particles similar to the electron, like for example muons. These charged particles can then if they have enough energy, emit light

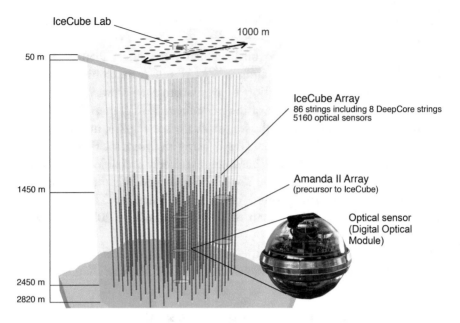

Fig. 3.6 The IceCube neutrino telescope. The individual sensors are lined up on strings, like pearls on a necklace, and lowered into holes drilled into the ice, ranging from 1.5 to 2.5 km below the surface of the ice. The strings are distributed over an area of roughly a square kilometre, hence the actual instrument or telescope has a volume of a cubic kilometre. The image of a single optical sensor (or Digital Optical Module) is at the lower right of the figure. Image credit: NSF/IceCube

known as "Cherenkov radiation", due to the same physical process that gives rise to the blue glow in the water basin of a nuclear reactor.[17]

Because neutrinos only interact very rarely with other particles, we have to give them very many opportunities for interaction. To detect neutrinos through their Cherenkov light, we have to build really big detectors containing exceedingly many water molecules.

One ingenious possibility to maximise the detector size, is to use the ice of the Antarctic as detector material. The IceCube Neutrino Observatory, or Ice-Cube does precisely this, it is a neutrino telescope or observatory constructed at the Amundsen–Scott South Pole Station in Antarctica, see Fig. 3.6. Completed in 2010 it is now taking data. IceCube consists of thousands of sensors that have

---

[17] Cherenkov radiation comes about when a charged particle travels through a medium with a speed higher than the speed of light of the medium (the speed of light in matter is smaller than the speed of light in vacuum).

been lowered into boreholes in the ice from the surface; they are distributed over a volume of one cubic kilometre 1.5 km under the Antarctic ice. The sensors, extremely sensitive light detectors called photo-multipliers, are lined up on "strings" or cables that also transmit the signals to the IceCube laboratory on the surface. When a neutrino interacts with a water molecule in the IceCube volume, the charged particle created in the process will leave a trail of light, allowing the individual detectors along its path to record its flight. From this the scientists in the laboratory can reconstruct the origin and speed of the original neutrino.

Despite its massive size and the mind-bogglingly large number of neutrinos passing the IceCube observatory every second, because they interact weakly and the detector is only sensitive to neutrinos of a particular energy (energetic enough to lead to Cherenkov radiation), only a couple of neutrinos actually get detected every year. For example from May 2010 to May 2012 only 37 neutrinos were observed at high enough energies for researchers to be confident that the neutrinos came from an astronomical distance away.

3.3.4.2   LIGO

There are several gravitational wave observatories taking data today, for example GEO600 in Germany, and VIRGO in Italy. Arguably the best known one at the moment is the "Laser Interferometer Gravitational-Wave Observatory" or LIGO. It is located in the USA and consists of two individual observatories, the LIGO Livingston Observatory in Louisiana, and the LIGO Hanford Observatory in Washington of similar design, roughly 3000 km apart.

The observatories are each made up of two large concrete tubes 4 km long, set at right angles in an L-shape, which meet in a laboratory building, as can be seen in Fig. 3.7; each LIGO observatory is designed to be a very large interferometer.

The basic principle of an interferometer is rather straightforward, as we show in a very simplified sketch of the LIGO observatory in Fig. 3.8. We first have to explain "interference" in this context. We start by considering two electromagnetic waves with the same amplitude, wavelength and frequency, for example monochromatic light. If we superimpose the waves so that the peaks of the waves coincide, the waves are "in phase", their amplitudes will add up, and we have "constructive interference". If the waves are out of phase, that is if the peaks aren't aligned exactly, the resultant wave will have an amplitude smaller than in the previous case. If the peaks and the troughs fall together, they will cancel each other out (one wave "going up" the other "going down")

Fig. 3.7 The LIGO gravitational wave observatory in Louisiana. The laboratory buildings are in the foreground, the 4 km long vacuum tubes are clearly visible in the vegetation. *Image credit: Caltech/MIT/LIGO Lab*

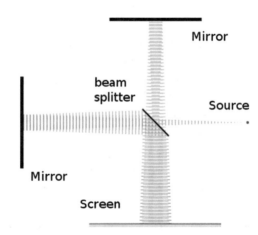

Fig. 3.8 A schematic interferometer to detect passing gravitational waves. Compared to Fig. 3.7, we would have the light source, usually a laser, the beam splitter and the screen (or detector) in the laboratory building, the "beams" in the vacuum tubes, and the mirrors housed at the end of the tubes

and we have "destructive interference". We note that larger amplitude means increased brightness.

Next we consider what happens to monochromatic light that starts out in phase as a single beam, but then gets divided up into different beams, and we let the beams travel along different paths. Depending on the length of the paths, the light in some of the beams will be out of phase, some other beams will be still in phase with each other. If at the end the light of all the beams is put back together again and directed onto a screen, we will end up with an interference pattern, that is brighter areas where the light is in phase, less bright ones where the light is out of phase, and dark areas where the light interferes destructively.

With this in mind, let us now return to the simple interferometer in Fig. 3.8. The laser provides us with light of a single colour—that is a *single* wavelength—that has the same phase, which means all wave trains oscillate in such a way that the wave crests are on top of each other (using techno-jargon, a laser is a "coherent light source"). In the beam splitter the laser beam, still in phase is split into two, one beam travelling down each tube, and getting reflected at the end of the tube at the mirror. The beams return and when they encounter the beam splitter again, the light is diverted onto a screen where we might see an interference pattern, since the tubes are not precisely the same length. We should stress again that this description is rather simplistic, but contains the essential physics.

The crucial point is that the interference pattern will change if the distance that light has to travel in one or both arm changes. This will happen when a gravitational wave passes through the interferometer and distorts spacetime, moving the mirrors by a tiny fraction depending on the direction of the passing gravitational wave. In this set up the mirrors act as "test masses" that are allowed to "ride the waves", similar to how a cork on the surface of a lake bobs up and down when a surface wave passes. Therefore the mirrors are very delicately connected to the tubes themselves and are allowed to "swing freely" from the tube ceiling. We will discuss the concept of spacetime in more detail in Sect. 6.4.2.2, and gravitational waves in Sect. 6.4.2.3.

Although the basic principle is simple, the technological execution is extremely difficult, since the distances involved are so small. After all, what the gravitational wave observatory is supposed to measure is changes in the distance the laser light travels which are a fraction the size of an atom!

Gravitational waves were observed directly for the first time in September 2015. Both LIGO observatories saw the distinctive change in the length of the tubes due to the merger of two black holes. The event was registered in both of LIGO's observatories, convincing scientists that this was a real event and not just some technical glitch, despite the smallness of the signal. The data taken by the two observatories is shown in Fig. 6.15.

3.4   Future Experiments and Telescopes

Above we highlighted some important telescopes, following our personal bias, that are taking data or have already been decommissioned. In the following we highlight some upcoming experiments, again choosing them by personal preference.[18] All these experiments have cosmology as one of their primary goals.

3.4.1 SKA

The Square Kilometre Array (SKA) is a giant radio telescope, consisting of two arrays of radio telescope dishes in Australia and South Africa. It is at present in advanced planning stage, with several prototype dishes already built in both countries. The dishes can be electronically connected to act as a single, giant radio telescope. The combined antenna area will be roughly one square Kilometre (hence the name).

The estimated costs for the SKA are at present 1.5 billion Euro. Construction started in earnest in 2018, with some precursor or prototype telescopes already taking data (these prototype telescopes are often referred to as "pathfinder"). Preliminary data collection of the SKA is to begin in the early 2020s.

3.4.2 Euclid

Euclid is a satellite mission at present being assembled by the European Space Agency. It has a 1.2 m diameter mirror on board and will observe in visible and infrared wavelengths. Euclid is named after the Greek mathematician Euclid,[19] who lived in Alexandria in the fourth century before our time.

The satellite will map the Large Scale Structure of the universe up to distances of several billion parsec, by observing roughly 10 billion astronomical sources. Researchers hope that this will help to explain the nature of dark energy and the geometry of spacetime. The launch date is currently set for 2022, with a mission cost set at roughly 0.5 billion Euro.

---

[18] We would like to apologise to our colleagues for the many omissions!

[19] Euclid of Alexandria (around 300 BC), Greek mathematician, major works attributed in geometry.

3.4.3 Extremely Large Telescope—ELT

The Extremely Large Telescope, or ELT, will be the largest optical telescope with a single mirror ever built, see Fig. 3.9. It should also receive a prize for least imaginative name.

The telescope will have a 39.2 m diameter primary mirror, observing in the visible and near infrared parts of the spectrum. It will be situated on Cerro Armazones, Chile, near Paranal Observatory (close to the Atacama desert). The estimated costs for the ELT are at present roughly 1 billion Euro. Building work began in June 2014, with preliminary data collection to begin in 2025. The primary mirror consists of 798 hexagonal segments, each measuring 1.45 m across and 50 mm thick. They can be moved individually to compensate for some deformation of the mirror segment due to changes in temperature and the influence of gravity. The telescope has a very complex optical system with overall five mirrors (including the primary). The fourth mirror is designed to "correct in real time for high order wave-front errors" that is for example atmospheric distortions from turbulence.

Fig. 3.9   The Extremely Large Telescope (ELT), with a mirror of 39.2 m in diameter it is indeed extremely large (note the size of the cars at the right of this computer graphic). The dome housing the telescope has a height of nearly 74 m and measures 86 m in diameter. *Image: ESO/L. Calçada*

3.5 Closing Comments

We have briefly described some of the key pieces of equipment used to collect observational data to study the universe, and in particular test and calibrate our cosmological models. The advances in theoretical cosmology were only possible, indeed often inspired and guided by new observational data sets. These new data could only be taken as new and ever more powerful telescopes and instruments became available.

We should stress that observations made using different instruments that observe at different wavelengths, or even use different messengers, are complementary. Observing a galaxy using a telescope that operates at visible wavelength and a radio telescope allows us to get different information about this galaxy, such as its shape and size but also which regions of the galaxy emit radio signals indicating for example the formation of new stars.

Finally we should highlight again the problem that we can only observe the cosmos—we cannot perform experiments "making different universes" and study these universes in a lab. But we can set up hypotheses that we can test using the observational data. If the hypothesis agrees with the data, the model remains valid, but subsequent observations may force us to reject the hypothesis.

4

What Do We Observe?

In the previous chapter we introduced and discussed, from the theoretical cosmologist's vantage point, the fantastic tools astronomers have at their disposal to make observations. Certainly astronomers do not just stare at the sky on a whim, so what *do* they actually observe? And what do they do with their observations once they are made? We will answer the last question at the end of this chapter. As we will discuss there, cosmologists are less interested in particular images of objects such as stars and galaxies but would like to have maps of the universe, the bigger the better. To make these maps we need to remember that our theory of the universe also informs the observations we make and how we interpret them. In this context in particular it affects the way we measure distances.

4.1 Distance Units in Cosmology

Although pictures of individual stars and galaxies are exciting and interesting in themselves, if we want to understand the universe on very large scales, then the distribution of these galaxies is even more important. To that end we have to measure the position of the objects we observe. To do this we first have to introduce the units appropriate for cosmology, and introduce some typical astronomical length scales, something we have omitted so far.

The distances and scales relevant for cosmology are vast, and cosmologists therefore usually use special units adapted to these large distances. Nevertheless the numbers involved are still mind boggling.

© Springer Nature Switzerland AG 2019
K. A. Malik, D. R. Matravers, *How Cosmologists Explain
the Universe to Friends and Family*, Astronomers' Universe,
https://doi.org/10.1007/978-3-030-32734-7_4

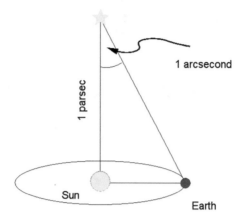

Fig. 4.1 The definition of a parsec: a star is at a distance of 1 pc from the Sun, if the angle at the top is 1 arcsecond. This is a "simple" triangulation using the diameter of Earth's orbit as the baseline

A standard distance unit in cosmology is the *parsec*, one parsec is equivalent to 3.09×10^{16} m. The definition is illustrated by Fig. 4.1: a star is 1 parsec away from the Sun, if we triangulate its position using the distance Earth—Sun as the baseline, then the angle at the top of the triangle star-Sun-Earth is 1 arcsecond (or 1/3600 of a degree).

Another common unit is the lightyear, defined as the distance light travels in vacuum within 1 year. The speed of light is a universal constant, and it travels 299,792 km/s in vacuum, hence a lightsecond is equivalent to 299,792 km, roughly the distance to the Moon from Earth. A lightyear is therefore equal to 9.4607×10^{15} m. To get at least a rough idea of what these numbers mean we use some familiar settings in our local neighbourhood as examples:

- the mean diameter of Earth is about 0.0425 lightseconds, in other words, it would take a light signal 0.0425 s to travel from the centre of the Earth to its surface,
- satellites in geostationary orbits used for communication are about 0.119 lightseconds or roughly 36,000 km from the surface of the Earth,[1]
- the distance from the Earth to the Moon is about 1.282 lightseconds,
- the average distance from the Earth to the Sun is 499 light-seconds, or 8.317 light-minutes, which means it takes roughly 8 min for an observer on Earth

---

[1]This will lead to a noticeable delay in a phone call across the Atlantic, though the "lag" in mobile phone networks usually has other causes.

to notice something happening on the sun (for example a solar flare going off),

- and finally, leaving our direct neighbourhood, the nearest star to the Sun, Proxima Centauri, is about 4.24 light-years or 1.3 pc away. This means for example it takes a light- or radio signal from the opening of the 2020 Olympics in Tokyo nearly 4 years and 3 months to reach Proxima Centauri and give the good news to the people there, if there are any.

Note that 1 pc is roughly equivalent to three lightyears. Although lightyears might be more familiar to the reader, the parsec is more commonly used in cosmology, and therefore we will often give distances in parsecs as well as in lightyears.

Before we discuss how the vast distances we are dealing with in cosmology can be measured, we will first introduce some of the objects that astronomers and cosmologists deal with. This also gives us the opportunity to show some of the amazing images the telescopes mentioned in the previous chapter have made. We focus on observations in the optical part of the electromagnetic spectrum, mainly for aesthetic reasons—these are, arguably, some of the prettiest pictures. However, there is much information also at other wavelengths.

4.2   Things We Can Observe Directly

We begin this section with a brief discussion of galaxies. A galaxy consists of stars, gas, and dark matter, which are held together by their own gravity. Galaxies come in all forms and sizes: small galaxies, usually referred to as "dwarf galaxies", might contain just tens of thousands of stars, whereas the largest galaxies have up to a hundred trillion, or 10^{14}, stars. The dark matter dominates the mass of a galaxy, the stars and the gas only contribute 5–10% of its overall mass. We will discuss dark matter in some detail in the next chapter.

Galaxies range from about 1000–100,000 parsecs in diameter in size. As discussed in the previous section, a parsec is a length unit popular in astronomy and roughly equivalent to 30 thousand billion kilometres.

The reader might have noticed that we also casually slipped "stars" into the definition of galaxies, without explaining what we mean by "star". For our purposes it is sufficient to define a star as a sphere consisting of mainly hydrogen and helium, which is held together by its own gravity. The pressure and the temperature at its centre is so large that hydrogen fusion starts, that is hydrogen atoms fuse together to form a helium atom, and large amounts of energy are released. A typical example of a star is our Sun. Very hot stars appear

Fig. 4.2 The Andromeda galaxy, as seen by the Hubble Space Telescope (HST). This an example of a spiral galaxy, similar to our own galaxy, the Milky Way. *Image credit: NASA/HST*

blueish, whereas cooler stars usually are more red. This is familiar effect, for example a "red hot" piece of steel is cooler than a "heated white" one. However, in many situations cosmologists do not care too much about objects as "small" as stars, or any objects smaller than galaxies or even clusters of galaxies.

We begin with an image of a typical spiral galaxy, the Andromeda galaxy, also known as M31, as observed by the HST, Fig. 4.2. The galaxy consists of billions or even trillions of stars, which are concentrated in the centre or bulge, and in the spiral arms stretching out from the bulge. The Milky Way, our own galaxy, is also a spiral galaxy and very similar to Andromeda. The Milky Way consists of roughly 300 billion stars, whereas Andromeda galaxy contains more than a trillion, that is 10^{12}, stars.

Another nice picture of a spiral galaxy called NGC 1232, is given in Fig. 4.3 (in the left panel). This image was made using the ground based Very Large Telescope (VLT) in Chile.[2] We round off our galaxy picture gallery with an image showing an edge-on view of a galaxy, again taken by the VLT, Fig. 4.3 (on the right). This galaxy is called NGC 4565.

Besides spiral galaxies, there are also other types, namely elliptical and irregular galaxies. They have very distinct shapes, as can be seen from the examples below in Fig. 4.4. All types play an important role in modern astronomy, and indeed the study of galaxies and how they form is a very important branch in

---

[2]The designation, or "name", of this galaxy stems from the acronym for "New Galaxy Catalogue", that is NGC.

Fig. 4.3 More spiral galaxies. On the left panel, the spiral galaxy NGC 1232, an image taken by the Very Large Telescope (VLT). This is a nice example of how spiral galaxies appear when seen from above or below. On the right panel the spiral galaxy NGC 4565, as seen by the VLT. This provides a nice example of how spiral galaxies appear when observed edge on. *Image credit: ESO*

Fig. 4.4 Examples for "non-spiral" galaxies. On the left panel, the giant elliptical galaxy M87, as seen by the Canada-France-Hawaii Telescope. On the right panel the irregular galaxy NGC 1427A, as seen by the VLT. *Image credit: ESO*

astronomy and astrophysics. However, from our cosmologist's point of view, it is sufficient to treat all types of galaxies equally.

As an example of an image not taken in the visible wavelengths we show, in Fig. 4.5, a picture of the spiral galaxy M51 (NGC 5194), also known as the "Whirlpool Galaxy" taken by the VLA radio telescope. The galaxy is at a distance of 7.9 million parsecs or roughly 24 million lightyears from Earth. The image was taken at a wavelength of 20.5 cm (or 1.46 GHz), and the

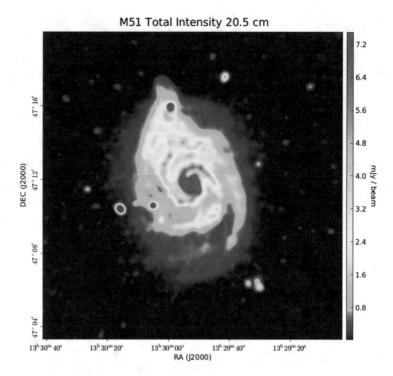

Fig. 4.5 The spiral galaxy M51 as "seen in radio waves". The image was taken by the VLA radio telescope at a wavelength of 20.5 cm (or 1.46 GHz). The colours indicate the total intensity emitted by the galaxy at this wavelength (red high intensity, blue low intensity). *Image Credit: MPIfR, Bonn*

colours indicate the total intensity of the radiation emitted by the galaxy at this wavelength. Red regions emit at high intensity, whereas blue regions emit at low intensity. These kind of observations allow astronomers to study for example the magnetic field of the galaxy.

The scales on which stars are formed are, at the moment at least, not particularly relevant to cosmology, as this happens on cosmologically very small scales. However, since it is such a beautiful image we have included another HST image here, the "Pillars of Creation" in the Eagle Nebula, Fig. 4.6.[3] The pillars are dense clouds of dust and gas, in which new stars form, measuring roughly 4 lightyears in length. The Eagle Nebula, also known

---

[3]In this context a "nebula" is a small, by cosmological standards, region of space containing gas, dust, and often a couple of stars.

Fig. 4.6 "Pillars of Creation" in the Eagle Nebula (or M16), another image taken by the HST. Dense clouds of dust and gas, the birth place of new stars. *Image credit: NASA/HST*

as M16 (in the Messier catalogue of astronomical objects) is part of the Milky Way, a distance of 6500 lightyears from Earth.

4.3 How to Measure Distances in Cosmology

In astronomy, as in every day life, if we want to know where things are, we have to determine the positions of these things. Without spending too much thought, we usually refer to a point of origin and speak about whether an object is, for example, above, to the left and in front of something. Speaking more abstractly, we need three numbers to determine the position of an object in space. It would be ideal if these numbers or coordinates would be simply attached to each object we observe. Unfortunately this is not the case and in cosmology the task of determining the correct position of an object is made more complicated due to the vast distances involved. Indeed, astronomers have to go to great lengths to determine the coordinates of distant objects.

To this end astronomers usually measure directions and the radial distance, that is the straight line distance, to an object. Measuring the direction is relatively simple, from the direction the telescope is pointing (we can use two angles, say). However, measuring the distance is much harder, as unfortunately this is not directly measurable.

In every day life we can use a simple tape measure to get the radial distance. In astronomy and cosmology this is unfortunately not possible.

Direct measurement, for example through sending radio signals and letting them bounce off an object and then measuring their time of travel is only technically feasible within the solar system. Already in the outer solar system the returning signal is too weak to be picked up on Earth. However, this method allowed astronomers to very accurately measure for example the distance from Earth to the Sun.

In the absence of direct measuring methods, astronomers have to resort to a system of indirect methods to measure distances up to cosmological scales. These methods are usually referred to as the "cosmic distance ladder", and each rung is used to calibrate the next higher one.

The first method, or the lowest rung of the ladder, used to determine distances beyond the solar system is also used on Earth: triangulation. From the endpoints of a baseline we measure the angle at which the object appears, see Fig. 4.7. This allows us then to calculate the distance to the object using simple geometry, or trigonometry to be specific. Since the distances in astronomy are so large, the angle will be very small and difficult to measure. To improve the accuracy of this method we can increase the baseline, and to make it useful

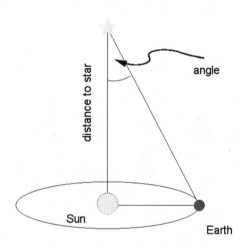

Fig. 4.7 The parallax method, which we already encountered defining the length unit parsec in the previous section, Fig. 4.1. We can use Earth's orbit around the Sun as the baseline for a triangulation. We know the length of baseline, the diameter of Earth's orbit, and from measuring the angle at which the star appears we can calculate the distance to the star

in astronomy we use the largest baseline available to us, that is the diameter of Earth's orbit around the Sun. This means we have to measure the angle at which the star appears twice, separated by 6 months. The parallax is then the angle at the tip of the triangle (it is actually defined as the half of this angle, which makes the distance Earth-Sun the baseline). With this method we can measure the distance to the nearest stars. For example, as mentioned above, the closest star in the solar neighbourhood is Proxima Centauri at a distance of 4.2 lightyears, about 1.3 pc. This method works to distances up to roughly 100 parsecs. The parallax method is closely related to the definition of the distance unit parsec encountered in Sect. 4.1. Measuring the parallax of a star is a direct method to determine the distance, it is nevertheless the first rung of the cosmic distance ladder.

For the next rungs astronomers rely on "standard candles" to measure distances. A standard candle is an object whose absolute brightness or luminosity is known and whose apparent brightness or luminosity can be measured. The underlying idea is that objects will appear less bright the further away they are. The method is readily explained using an example. Let's consider a 100 W light bulb. We know how bright it is, that is how much energy per second it emits, since by definition it is a 100 W light bulb.

The further away we walk from the light source, the light bulb, the dimmer it will appear. The absolute brightness of the source hasn't changed, it is after all still a 100 W bulb, however its apparent brightness has diminished as the light gets spread out over a larger area the further away we walk.[4] Therefore, from the apparent brightness, the amount of light that reaches us, we can calculate the distance to the object if we know its actual or absolute luminosity. The problem here is, to find the equivalent of 100 W bulbs in the universe, that is objects of known absolute luminosity.

Besides having a known absolute brightness or luminosity, a good standard candle should satisfy additional "requests" from astronomers. First of all, besides having a known luminosity or brightness, they should also be fairly bright, so that they can be seen out to large distances (an actual 100 W bulb wouldn't be seen from very far away). The objects should also be abundant, that is they should conveniently be in all the places and regions to which we would like to know the distance. Finally, the objects should be well understood,

---

[4]The light bulb or source emits light in all directions. Hence all the light from the bulb that passes through the surface of a spherical shell at 1 m distance from the bulb spreads out and also has to pass through a larger shell further away. Since the area of these shells will be proportional to the distance squared, the light intensity will decrease by the inverse of the distance squared. Hence light from a bulb ten times further away will be a hundred times dimmer, for example.

that is physicists should understand how these objects work. If for example the object were to get dimmer with time (like some old light bulb might), or its luminosity depends on its location (where the bulb was made), then this might introduce errors if astronomers are not aware of these possible problems.

There are many types of standard candles used in astronomy. We now look at two particularly popular examples of standard candles, Cepheid variable stars, and a particular type of supernova, namely type Ia supernovae.

Cepheids are stars that change their brightness periodically. What makes them so useful in astronomy or more particularly in cosmology is that they have a well defined period-luminosity relation, discovered by Henrietta Swan Leavitt.[5] This means that by measuring the period of the brightness fluctuations astronomers can deduce the absolute brightness of the Cepheid. The period of the fluctuations is usually of the order of tens of days. Astronomers use Cepheid variable stars as a method to determine distances out to tens of Megaparsec.

Recently supernovae have become popular as standard candles to measure cosmic distances. A supernova is an exploding star, but only a particular type of supernova, referred to as "supernovae of type I a" or SN Ia, can be used as standard candles.

This type of supernova consists originally of a binary star system, a white dwarf and a companion star. The white dwarf accretes or sucks material from the companion, until it reaches the so called Chandrashekar limit of roughly 1.44 solar masses. Reaching this limit the white dwarf implodes, because the pressure of the stellar matter in the core of the white dwarf can not support the gravitational force due to the outer layers of the star any longer, and the white dwarf begins to collapse. However, due to the collapse a new fusion reaction starts in the white dwarf and the star then explodes. Since this scenario always follows the same sequence of events, the brightness is always the same, roughly five billion times brighter than the Sun. Therefore a supernova of type Ia is indeed a bit like a 100 W bulb, but much, much brighter. Due to the their incredible brightness, supernovae of type Ia can be used to measure distances method up to Gigaparsec scales.

What do we gain from observing Cepheids and supernovae out to these incredible distances? We should stress, that in both of these cases, or for any other standard candle, what we usually really want to know is the distance to the galaxy or galaxy cluster that contains the standard candle. Standard candles

---

[5] Henrietta Swan Leavitt (1868–1921), American astronomer, discovered the period-luminosity relation of Cepheid stars.

are therefore essential tools to measure distances in cosmology. The use of super novae of type Ia as standard candles is discussed in more detail in Sect. 8.2.1.

In the 1920s Edwin Hubble,[6] building on work by Vesto Slipher,[7] used Cepheids as standard candles to measure the distances to nearby galaxies. He discovered that the light of the galaxies is redshifted, that is the wavelength of the light was stretched. In particular he found that the further away a galaxy is, the more redshifted the light we receive is. How could Hubble conclude that the light from distant galaxies is stretched, and what does "stretching of light" actually mean?

When light or electromagnetic radiation in general passes through a gas, the radiation can interact with the atoms and molecules that constitute the gas. During the interaction an electron in one of the atoms can absorb a photon, and it becomes "excited", that is the electron reaches a higher energy level, if it was originally in the right state.[8] After a very short time the electron releases the energy it gained by absorbing the photon, by emitting another photon but not necessarily at the same wavelength and in the same direction as the original photon. The overall net effect is that when radiation of a broad range of wavelengths, that is with a continuous spectrum, passes through the gas, there will be wavelengths afterwards where there is no radiation emitted. This leads to features, namely black absorption lines at well defined wavelengths, in the spectrum we observe. The position of these black line depends on the atoms in the gas that did the absorbing, and the energetic state these atoms were in at the time.

The light and the radiation we observe from a star is produced by the fusion of hydrogen into helium in the stellar core. It therefore has to pass through the outer layers of the star, before it can start its journey towards us. We therefore do not observe a continuous spectrum, but a spectrum with black absorption lines. The position of these lines allows astronomers to deduce from millions of lightyears away what elements there are in the star's atmosphere and in what state.

These processes also take place for example in the Sun's atmosphere. Electromagnetic radiation produced by the fusion of hydrogen into helium in

---

[6] Edwin Powell Hubble (1889–1953), American astronomer, worked in observational cosmology and discovered the redshift-distance relation.

[7] Vesto Melvin Slipher (1875–1969), American astronomer, discovered that spectral lines in spectra of galaxies were shifted.

[8] Here it is easier to think of the radiation in terms of particles, called photons. The photon gets absorbed by the electron, which in turn reaches a higher energy level. We shall revisit both atoms and the wave–particle issue later on in Sects. 5.1.1 and 5.1.4, respectively.

Fig. 4.8 *Middle:* a "wave train", either a sound wave or an electromagnetic wave, as seen by an observer at rest relative to the source, indicated by the star. *Top:* the wave train if the source and observer move towards each other, the peaks are bunched together compared to the wave seen by an observer at rest. *Bottom:* the wave train if the source and observer recede from each other, the distance between the peaks gets stretched, compared to the wave in the middle

the core of the Sun gets absorbed and re-emitted many times until it reached the surface of the Sun, which we then can see and observe.[9]

Let us now return to Hubble's discovery. Since galaxies are made of stars, the light that we observe from them will also have similar absorption lines in their spectra, the combination from millions of stellar spectra. However, the spectra that Hubble observed from far away galaxies were "wrong": although the absorption line pattern was as expected, the lines were in general shifted to longer wavelengths, longer than what we would expect from experiments in the laboratory or from observations of nearby stars. Since red light has longer wavelengths than for example blue light, as discussed in Sect. 3.2.1 above, this effect is called "redshift".

What is the origin of this redshift of the absorption lines in the galaxy spectra? There are several different effects contributing to the redshift of the light emitted by a galaxy, the most important ones here are due to the Doppler shift of the radiation, and due to the expansion of space itself.

The first effect, the Doppler shift, is very likely to be familiar to the reader in the case of sound waves. It is the same effect that makes the siren of a police car sound higher in pitch (higher frequency) when it approaches an observer, and lower pitched (lower frequency) when it recedes from the observer, compared to the sound when the police car is at rest with respect to the observer, see Fig. 4.8.

The same effect also changes the wavelength of light if a source either approaches or recedes from us. In the first case the electromagnetic "wave train"

---

[9]The element *helium* was first identified by its distinctive absorption lines in the light received from the Sun, even before it was "discovered" on Earth.

arriving at the observer appears to be compressed, compared to the original "wave train". If the source recedes, the light arriving at the observer is stretched out. Since shorter wave lengths are bluer, and longer ones more red, the name redshift for this stretching of light was introduced, see again Fig. 4.8.

The other contribution to the redshift (or blueshift) of electromagnetic radiation and therefore also the light emitted by a far away galaxy, is due to the expansion (or contraction) of space itself, as we will discuss in more detail in Sect. 6.4.2.4.

Hubble could therefore establish a clear redshift to distance relation. This means that if we measure the redshift at which an object such as a galaxy emitted light, we can calculate the distance to the object. This calculation is mildly complicated and requires knowledge of the matter constituents of the universe, and also assumptions on how it evolves. We will discuss this in the next two chapters. Hubble's law, or the redshift-distance relation, is therefore sometimes regarded as the last rung in the cosmic distance ladder. Hubble's law is used to measure distances on the very largest scales from tens of Megaparsecs to several Gigaparsecs.

Large surveys of galaxies use redshift to measure the distance of the galaxies from us. See for example Fig. 1.4 in the introduction, a map of galaxies made by the 2dF survey. On the right-hand-side of the figure we see at the top the distance given in redshift, and at the bottom the distance in billions of lightyears. The actual distance in parsecs or lightyears has to be calculated and depends on the underlying theory of gravity and the matter content of the universe, as we will discuss later in Sects. 6.4.2.4 and 6.4.4. Cosmologists therefore often do not bother to translate redshift into distance and instead simply give the redshift of the object, as this is a directly measurable quantity.

As pointed out above, these different observational techniques to measure distances constitute the cosmic distance ladder, and a method on a lower rung can be used to calibrate the method on the next higher rung. This interdependency of the different rungs was recently demonstrated again, when the period-luminosity relation of Cepheids was re-calibrated using the latest observations. Using this improved calibration the astronomers undertaking the observations could correct the distance from us to the Large Magellanic Cloud to 48 kpc, 2 kpc less than was assumed before. Needless to say, the actual position of the Large Magellanic Cloud had not changed by 2 kpc, the change was due to an improvement in our "tape measure", that is our measuring device.

With the exception of the lowest rung method, parallax or triangulation, these methods do not give distances directly. What we observe and what is directly measurable is the brightness (or luminosity) of an object, or on large scales the redshift at which the radiation is emitted. We then have to calculate the distances, using cosmology.

4.4 Putting Things into Perspective

In the previous section we have discussed very briefly how to measure astronomical distances. This allows us now to assign numbers to the objects in the images discussed in Sect. 4.3, putting them into context and relate them to each other. We remind ourselves and the reader that 1 pc is roughly 3.26 lightyears.

A "typical" spiral galaxy like our own, the Milky Way, similar in shape to the galaxies depicted above, has a diameter of roughly 100 thousand lightyears (about 35 kpc). The Pillars of Creation, Fig. 4.6, are inside the Milky Way and are 6500 lightyears away from Earth. The length of the pillars is roughly 4 lightyears. The Andromeda galaxy, our closest neighbouring galaxy, is roughly 2.5 million lightyears or 785 kpc away from the Milky Way, and has a diameter of roughly 220 thousand lightyears (or 67 kpc). Another galaxy of interest, NGC 1232, is 100 million lightyears away, and galaxy NGC 4565 at a distance of 30 million lightyears.

This gives us an idea how far we have to travel to find typical examples of different types of objects, or how they are distributed in space, and typical sizes. Stars within galaxies are separated from each other roughly on the order of parsecs, whereas galaxies are roughly Megaparsecs from each other, and have diameters of the order of tens of kiloparsecs.

In cosmology we are interested in the very largest scales. We have already introduced a map showing the distribution of galaxies on very large scales up to billions of parsecs by the 2dF galaxy survey, Fig. 1.4 in Chap. 1. Each dot corresponds to a galaxy, its position on the sky and the distance from us have been measured with the methods described in the previous section.

We can now discuss the map, Fig. 1.4 in some more detail. For clarity we add here a magnification of the central part on the right of Figs. 1.4 and 4.9 which shows the distribution of galaxies out to 1 billion lightyears (or roughly 300 Mpc). As mentioned above, the observer is at the centre of Fig. 1.4 looking outwards, or at the bottom of Fig. 4.9 looking up. The slices are roughly 75° wide, the left slice is about 7.5 and the right slice about 15° thick (the Northern and the Southern band, respectively). The survey can "see" galaxies up to distances of more than 2 billion lightyears or nearly a billion parsecs (corresponding to a redshift of about 0.2). Beyond that distance galaxies are too faint to be seen by the telescope. We should stress that this doesn't mean

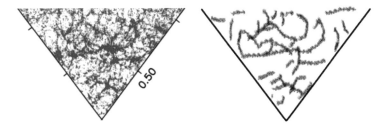

Fig. 4.9 The distribution of galaxies on large scales out to about a billion lightyears (or roughly 300 Mpc). The figure on the left is a zoom into the right slice of Fig. 1.4 in Chap. 1, from the 2dF galaxy survey. Each dot corresponds to a galaxy. At these scales we can see that the largest structures are at scales of up to about 100 Mpc. On these scales the filamentary structure in the galaxy distribution becomes apparent, which is often referred to as the "cosmic web". On the right, we highlighted the densest regions, to emphasise the web-like structure in the distribution of galaxies

that beyond this distance there are no galaxies! Bigger telescopes and longer observing times will extend these surveys. Another means to extend these surveys is to use particularly bright galaxies, called quasars, for the survey, as we will briefly discuss at the end of this section.

From Figs. 1.4 and 4.9 we see that the universe on its very largest scales is smooth and more or less featureless. There are very large structures up to several hundred million lightyears in size or roughly of the order of 100 Megaparsecs, but no larger structures. We will discuss the implications of this observational fact later on in Sect. 6.4.2.4. The 2dF survey measured the redshift and the position on the sky of hundreds of thousands of objects in two "pie-shaped" slices. The distance to the objects can then be calculated from the redshift. The survey covered an area on the sky of about 1500 square degrees. The name "2dF" derives from the fact that the survey instrument covers an area of approximately two square degrees at a time. In total the 2dF survey measured the photometry, or the light intensity at different wavelengths, of 382,323 objects which included spectra for 245,591 objects, of which 232,155 were galaxies (221,414 with good quality spectra), 12,311 stars, and 125 quasi-stellar objects or quasars. The survey necessitated 272 required nights of observation, spread over 5 years.

Figures 1.4 and 4.9 show the distribution of galaxies on very large scales. To get an idea of what our galactic neighbourhood looks like, we can use the same methods to measure distances that we used to make these maps.

The galaxies closest to our own galaxy, the Milky Way, are depicted in Fig. 4.10. The Milky Way is part of what is known as the "local group" of galaxies, which includes the Andromeda galaxy, shown in Fig. 4.2. Roughly 50

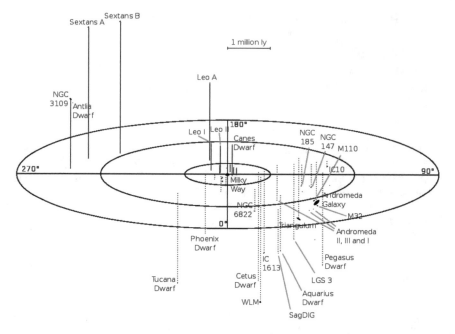

Fig. 4.10 Our local group of galaxies, the Milky Way is at the centre of the figure. Roughly 50 galaxies belong to the local group, including for example the Andromeda galaxy, which is at a distance of roughly 2.5 million lightyears, or 0.8 Mpc, from us. The local group measures roughly 10 million lightyears in diameter. *Image credit: Richard Powell*

galaxies belong to the local group, which measures about 10 millions lightyears, or a couple of Megaparsec, across.

Our local group of galaxies is itself part of the Virgo super-cluster of galaxies, as can be seen in Fig. 4.11. The Virgo super-cluster contains roughly 100 galaxy clusters and has a diameter of about 30 Mpc, roughly hundreds of million lightyears.

In 2014 astronomers in Hawaii discovered that the Virgo cluster seems to be part of an even larger structure, the Laniakea Supercluster.[10] This supercluster contains roughly 100,000 galaxies and is about 150 Mpc or 500 million lightyears large.

So far structures like the Laniakea Supercluster seem to be roughly the biggest structures, or to be precise, gravitationally bound structures in the universe. No structures larger than of the order of hundred Megaparsec have

---

[10]The name "laniakea" is Hawaiian and can be translated as "immense heaven".

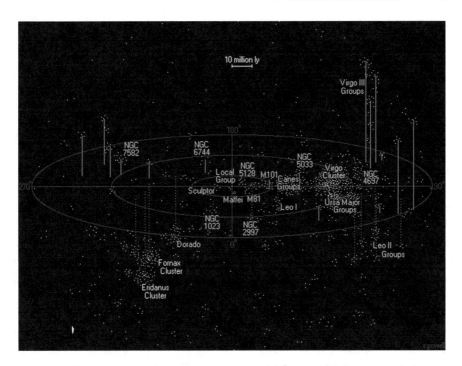

Fig. 4.11 The Virgo super-cluster containing roughly 100 galaxy clusters, including the local group, which is at the centre of the figure. The Virgo super-cluster has a diameter of roughly 30 million parsec. *Image credit: Richard Powell*

been observed. This is in agreement with our current understanding of how structure in the universe forms, namely by smaller structures assembling through gravitational attraction into larger structures. This process takes time, and since the beginning of the universe no larger structures than the ones described "had time" to form.

Even larger than the 2dF survey is the Sloan Digital Sky Survey or SDSS. The SDSS uses a dedicated 2.5 m telescope to systematically map a large section of the sky and a similar survey, the 2dF survey, used the 3.9 m Anglo-Australian Telescope. It is safe to say that most modern telescopes do not simply "take pictures" of the objects under study, but also can record additional information on the objects, usually in the form of spectra.[11] What made these telescopes stand out at the time of their inception is their ability to rapidly collect and process data from many objects in an image. The SDSS survey

---

[11]The spectrum provides information on how much radiation or light is emitted by an object at a particular wavelength. This is often also referred to as spectrometry.

began in 2000, and has so far mapped over 35% of the sky, with photometric observations, measuring the brightness, of around 500 million objects and spectra for more than 1 million objects. The SDSS covers an area on the sky of nearly 15,000 square degrees.

The SDSS telescope uses a *drift scanning technique*, which keeps the telescope fixed and makes use of the Earth's rotation to record small strips of the sky. The image of the stars in the focal plane drifts along the CCD chip, instead of staying fixed as in tracking telescopes. This method allows consistent astrometry over the widest possible field and precision remains unaffected by telescope tracking errors. Every night the telescope produces about 200 gigabyte of data.

Figure 4.12 shows the distribution of galaxies as measured by the SDSS, out to distances of 1.5 billion lightyears or 500 Mpc from Earth. As in Figs. 1.4 and 4.9, we see that galaxies are distributed in a web-like structure on large scales, with the largest structures extending to roughly hundreds of Megaparsec.

The SDSS will run until 2020 and is mapping galaxies out to distances of about 7 billion light years, or roughly 2 Gigaparsec, beyond this distance galaxies are too faint to be observed by the survey's telescope. To map the large scale structure beyond this distance, SDSS also maps the distribution of "quasars".

Quasars, short for "quasi stellar objects" are a particular type of galaxy with a super-massive black hole at their centre weighing millions of solar masses, surrounded by a large disk of matter called an accretion disk. Large amounts of material from the accretion disk gets sucked towards the black hole, as the matter spiral inwards on smaller and smaller orbits. The material heats up through friction to such an extent that it emits huge amounts of electromagnetic radiation which allow astronomers to observe quasars at radio, infrared, visible, ultraviolet, and X-ray wavelengths.

When first observed, quasars were mistaken for point-like objects, just like stars, because they are so far away. They are extremely bright and hence visible up to very large distances of billions of parsec. The furthest quasar observed at the moment (2019) is at a redshift of 7.54, corresponding to a distance of 9 Gigaparsecs or 29 billion lightyears from us. The object was named ULAS J1342+0928, and emitted its light roughly 690 million years after the beginning, meaning the light took 13.1 billion years to reach us. For us to observe the quasar across this vast distance it has to be extremely bright: its luminosity is equivalent to 4×10^{13} solar luminosities, that is ULAS J1342+0928 is as bright as 40,000 billion suns.

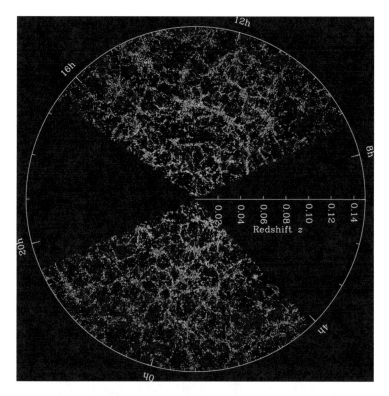

Fig. 4.12 The distribution of galaxies as seen by the Sloan Digital Sky Survey (SDSS). The SDSS is similar in kind to the 2dF survey, but observing different and larger areas of the sky, measuring the positions of many more objects. The outer edges of the map are roughly at a distance of 1.5 billion lightyears or 500 Mpc from us. We and the telescope are the centre of the map looking outwards, leading to the distinctive "pie shape" again. *Image Credit: M. Blanton and SDSS*

To make even larger maps of the large scale structure of the universe than possible using galaxies, the SDSS used quasars to extend these maps out to redshifts of about 3.5, as shown in Fig. 4.13. A redshift of 3.5 also corresponds to a look back time of 12 billion years, that means the light from quasars at this redshift travelled for 12 billion years to reach observers today. The distances involved are even more impressive, when expressing them in parsec and lightyears: a redshift of 3.5 corresponds to a distance (today) of 6.9 billion parsec or 20.7 billion lightyears.[12] The distance in lightyears is larger than just

---

[12]A redshift of 1100 corresponds to a distance (today) of 14 billion parsec or 42 billion lightyears. This is close to the size of the visible universe is today 14.2 billion lightyears or 42.6 billion lightyears, as we will see in Sect. 6.4.4.

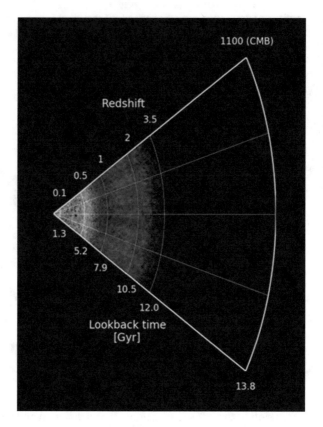

Fig. 4.13 An even larger map of large scale structure from the SDSS including quasars. This map is similar to Fig. 4.12, but we are on the left instead of in the centre. At the top of the slice redshift, at the bottom "look back time" as distance units. The yellow dots show the location of galaxies, as in the previous maps, but now the map also shows the position of quasars as red dots, extending the map to much larger distances. Quasars are much brighter and therefore visible out to much larger distances. Note, the cosmic "dark ages" range from about redshift 20 to 1100—the Cosmic Microwave Background. *Image Credit: Anand Raichoor (EPFL, Switzerland) and the SDSS collaboration*

the lookback time due to the expansion of the universe, which allows to receive information from further away, as we will discuss in Sect. 6.4.4. If the universe would not expand, the distance in lightyears and the look back time would be the same.

We also see in Fig. 4.13 that there are only very few light sources beyond this redshift. This is because at larger distances even quasars aren't bright enough to be observed with the current generation of telescopes (at the moment, the object observed at the furthest distance is the quasar discussed above at redshift 7.54). But beyond a redshift of about 20 there were no objects that emitted visible light, and therefore the region between redshift 20 and 1100 is called the cosmic "dark ages", as we will discuss in Sect. 8.2.2. At redshift 1100 we have on the right of Fig. 4.13 the Cosmic Microwave Background, which formed when the universe was 380,000 years "young". We will discuss its formation in Sect. 8.2.3, and will discuss the map of the Cosmic Microwave Background next.

4.5 An Even Larger Map: The Cosmic Microwave Background

We conclude this chapter with observations of the microwave sky and another map on even bigger scales that we already encountered in the introduction, Chap. 1, Fig. 1.3. Using the space- and ground-based microwave telescopes described in Sect. 3.3.3, observations of the Cosmic Microwave Background have become one of the most important sources of information for cosmologists.

The Cosmic Microwave Background is a remnant from the earliest epoch of the evolution of the universe and it offers a window into the physics of the early universe, a snapshot taken when the universe was just 380,000 years old. At this time the universe had cooled down to about 3000 K, enough for the electrons to combine with protons to form neutral hydrogen. This also meant that the electromagnetic radiation no longer interacted with the electrons, that is the radiation and the electrons decoupled and this time is therefore also known as the time of "decoupling". The Cosmic Microwave Background is therefore the oldest "object" we can see, also sometimes referred to as the "oldest light" in the universe. The electromagnetic radiation could, from the time of decoupling, travel without any further interruption to astronomers and their telescopes today—if the radiation didn't hit any objects on its way. We will discuss the general physics behind these processes in more detail in the next two chapters, and the formation of the Cosmic Microwave Background in Sect. 8.2.3. Here we restrict ourselves to describe the observational data that we have from this early time.

At the time when the radiation started its journey its spectrum peaked in the near-infrared, close to visible wavelengths. It resembled visible light as emitted by an object, a black body, with roughly 3000 K. Due to the expansion of the universe the electromagnetic radiation got stretched to longer wavelengths, and now peaks in the microwave region of the spectrum, corresponding to a black body with just 2.73 K or roughly $-270\,^{\circ}$C.

We can now ask, what would we observe if we look at the sky with our microwave telescopes? The first thing we notice is the microwave radiation coming from our own galaxy. Luckily this is not the main contribution at all microwave wavelengths, and in some wavelengths or frequencies we do observe the Cosmic Microwave Background. Once we have taken into account the distortion of the Cosmic Microwave Background due to the movement of Earth (actually Earth and the Milky Way) relative to the Cosmic Microwave Background, we observe that the Cosmic Microwave Background radiation is very uniform, and as we mentioned previously has today on average a temperature of 2.73 K. However, the Cosmic Microwave Background temperature is not perfectly uniform, it has fluctuations at the level of 1 part in 100,000, or 10^{-5}. These fluctuations in the Cosmic Microwave Background, also referred to Cosmic Microwave Background anisotropies,[13] can be measured despite their smallness at the micro-kelvin level. Besides measuring these temperature fluctuations, which is the same as measuring the fluctuations in the intensity of the Cosmic Microwave Background radiation, we can also measure the polarisation of the radiation. If the electromagnetic radiation or waves are polarised, they have a preferred plane of oscillation: instead of all electric fields of the individual waves oscillating in random directions, they have overall a preferred direction (the same holds for the magnetic fields, as they are always at right angles to electric fields, see Fig. 3.1). We will discuss the underlying physical process in the following chapters.

Finally to conclude this very brief discussion of the observational aspects of the Cosmic Microwave Background, we present a map of the Cosmic Microwave Background anisotropies as seen by the Planck satellite (in 2015) in Fig. 4.14. The map shows a projection of the whole sky, and is essentially a snap shot of temperature fluctuations, the hot and cold spots, at the time of decoupling. The average size of the fluctuations on the sky is roughly 1°, in comparison the angular size of the moon is roughly 0.5°. The average size of the fluctuations corresponds to 145 thousand parsec at the time of decoupling, that is when the Cosmic Microwave Background photons started their journey, or 160 Mpc today, due to expansion of the universe.

The temperature fluctuations in the Cosmic Microwave Background can be related to the earliest epochs of the universe, even before the time of decoupling. How the pattern of temperature fluctuations shown in Fig. 4.14

---

[13]The temperature of the Cosmic Microwave Background is on very large scales isotropic, that is the same in all directions. "Anisotropies" are small direction-dependent deviations from this uniform temperature.

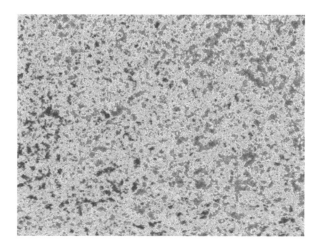

Fig. 4.14 Zooming into Fig. 1.3 of Chap. 1, an area measuring 20 by 30° on the sky: a map of the small temperature fluctuations in the Cosmic Microwave Background as seen by the Planck satellite. The red areas are warmer and the blue ones colder than the average temperature, ranging from −0.0003 to 0.0003 K around the average of 2.725 K (or about −270 °C). A typical hot or cold "spot" has an angular size of roughly 1° on the sky. *Image credit: ESA/Planck*

comes about will be discussed in Chap. 7 and Sect. 8.2.3. First we have to take a closer look at the constituents of the universe in the next chapter.

It is nice to know in this age of flat screen LED TV sets, that the Cosmic Microwave Background contributes a little to the "snow" or static in old cathode ray TV sets. Hence "in the olden days" after the end of broadcast everybody became a cosmologist.

5

What is the Universe Made Of?

If we want to construct a model of the universe we need to know what is in it and we also need to know the physical properties of whatever it is. In this chapter we will therefore have to answer the questions: what stuff, what kinds of matter, is there and where does it come from. Of course we are interested in the "stuff" from the point of view of cosmology and building our cosmological models. This means that we gloss over or even leave out some stuff that is important in other branches of physics.

5.1 Familiar Types of Matter, What Is Normal Stuff Made Up Of?

We have mentioned 'in passing' some of the matter constituents in the previous chapters but will now discuss the different types of matter or stuff, relevant for cosmology, in more detail.[1]

5.1.1 Matter on the Smallest Scales: Up to the Size of Atoms

Before looking at the properties of matter on large, cosmological or even on everyday "human" scales it is necessary to take a closer look at the matter on

---

[1] We use the terms "matter" and "stuff" interchangeably.

© Springer Nature Switzerland AG 2019
K. A. Malik, D. R. Matravers, *How Cosmologists Explain the Universe to Friends and Family*, Astronomers' Universe, https://doi.org/10.1007/978-3-030-32734-7_5

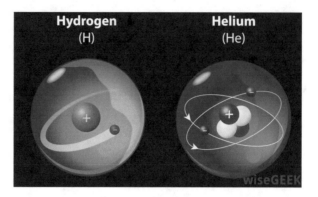

Fig. 5.1 The simplest atoms, hydrogen and helium. The hydrogen atom, on the left, consists of only a single proton, the hydrogen nucleus, surrounded by a single electron. The helium atom has a slightly more complicated nucleus consisting of two protons and two neutrons, which is surrounded by two electrons. *Image credit:* wisegeek.com

very small scales, which we will do after giving some basic definitions. The concept of "matter" is used throughout cosmology despite the fact that it is not defined uniquely—its meaning often depends on the context in which it is being used. The simplest, most basic, description is that matter occupies space, and has a rest mass. By rest mass we mean the mass an observer measures who is at rest relative to the particle, in agreement with what we might consider as mass in everyday life.[2] Furthermore, matter is made up of individual particles, and these particles can and might interact with each other.

Taking a less abstract view let us travel from macroscopic, every day scales to smaller and smaller scales. We would find that simple, normal matter consists of molecules, which are formed from chemical elements or atoms. Individual atoms are themselves made up of a nucleus, or core, surrounded by electrons, see Fig. 5.1. The nucleus itself is made up of protons and neutrons. Neutrons are charge neutral, whereas protons are electrically positively charged, and electrons negatively. The charges of the electron and the proton are equal and opposite. In electrically neutral atoms the number of protons is therefore balanced by the number of electrons. We will discuss the forces that act on the particles, for example attracting protons to electrons or holding nuclei together, in Chap. 6.

To get a rather crude picture of an atom we might imagine the electrons whizzing around the nucleus on tiny orbits, see Fig. 5.1. The actual behaviour

---

[2]For example, we can think of the "observer at rest relative to the particle" as a scientist measuring the mass of the particle in a laboratory.

of the electrons is more complicated, and we shall touch upon this in the next chapter. The chemical elements, and therefore the atoms, can be arranged in the periodic table, depending on the number of electrons in the outer "regions" of the atom, and the number protons and neutrons in the nucleus. To illustrate the structure, let us look at two examples, hydrogen and helium.

Hydrogen is the simplest atom, its nucleus or core consists of only a single proton, which is accompanied or orbited by a single electron, on the right of Fig. 5.1. The next element is helium. The helium atom has a slightly more complicated nucleus consisting of two protons and two neutrons, and therefore two electrons orbit the nucleus, see the left of Fig. 5.1.

Whereas the electron is an elementary particle, which means it has no further substructure, protons and neutrons are not and have further constituent components. They each consist of three quarks, which are however elementary particles. Quarks and electrons are therefore the most basic building blocks of everyday matter relevant for cosmology. To avoid justified complaints by particle physicists we should point out that there are many more subatomic particles than just quarks and electrons, but they play at the moment only a very limited role in the cosmology we are interested in. Also, although in the true particle physics sense only particles made up of quarks count as baryons, cosmologists often refer to all standard matter as baryonic matter (ignoring the fact that for example electrons aren't actually baryons).

The size of a proton is roughly 10^{-15} m, and the size of an electron is even smaller at 10^{-22} m. Using our simple atomic model above and imagining atoms as being small spheres, we find that the hydrogen atom has a radius of about 0.12 nm or 1.2×10^{-10} m, and the helium atom a radius of 0.14 nm. Hence the typical size of an atom is of the order of 10^{-10} m or a tenth of a nanometre. It is therefore fair to say that atoms consists mainly of empty space. This is despite in most circumstances behaving like or at least being well described as tiny, solid spheres. Later on we will see that also on the very largest scales the universe is rather empty.

Above we have described the matter constituents but haven't explained what keeps matter from falling apart, that is, what forces are acting on these tiny scales. Electrons are negatively charged particles, whereas protons carry a positive charge and neutrons are charge neutral. The negatively charged electrons are coupled to the positively charged nucleus, consisting of neutrons and protons, through the electromagnetic force. The protons and neutrons in the nucleus of an atom are held/bound together by the nuclear force. The protons and neutrons themselves consist of quarks, as mentioned above, and are held together by the strong force which acts between them. We will discuss forces in more detail in the next chapter.

5.1.2 Matter on Larger Scales: Many Particles

We will now use the rather simplistic view of particles as tiny billiard balls or spheres, interacting and bumping into each other, to explain the behaviour of matter on larger scales. Although not entirely accurate on atomic scales, on slightly larger scales this picture works surprisingly well.

From everyday experience we know that matter can take various states or phases—it can be a solid, a liquid, a gas or a plasma, the latter form probably being slightly less familiar. Let us use water as an example for the first three states. At standard pressure, water is a solid below $0\,°C$, it is in liquid form between zero and $100°$, and above $100°$ it is in its gaseous phase. We notice that we can get from one state to the next by "increasing the energy of the system", see Fig. 5.2. A simple way to increase the energy would be to heat the system. To explain this behaviour we resort to a naive version of the *"kinetic theory of matter"*. Let us assume that all particles undergo some random motion, which we refer to as "thermal motion", depending on the temperature of their surroundings; we mean by "particle" in this context the smallest constituent of the stuff we are considering, a water molecule in the case of water, or an electron in the case of an electron gas. The higher the temperature, the more the particles move around. At zero absolute temperature, the particles are at rest (at least very nearly). The higher the temperature, the faster the particles move around. As pointed out, this is a simplistic explanation, for example at

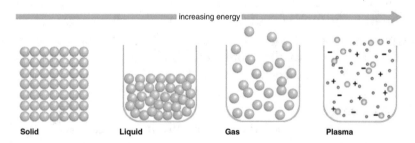

Fig. 5.2 The different states of matter, the energy of the system increases from left to right. On the left the matter is a *solid*, the individual particles (atoms or molecules) form a rigid structure. When we increase the energy of the system of particles, for example by heating, the rigid structure breaks up and we have a *liquid*, in which the particles can move easily relative to each other, but are still close together. Increasing the energy further, the particles form a *gas*, they move freely around in the container and aren't longer close to each other. Adding more energy we eventually have the matter in form of a *plasma*, the interactions so energetic that the particles loose some of their electrons. *Image credit: Encyclopedia Britannica*

low temperatures the particles are not totally free to move, but rather wiggle about (or oscillate) their positions.

We now return to the different states of matter, starting at low temperatures, see the image on the left of Fig. 5.2. The matter behaves like a solid, with the particles tightly packed together, they are held together by electromagnetic forces acting between the electrons that form the outer layers of the atoms. The particle speeds due to the thermal motion are small, the kinetic or thermal energy of the atoms is not high enough to break these bonds between the atoms. We might think of these bonds as little springs connecting each atom to its neighbours, allowing it to wiggle around but keeping it more or less in its place. It therefore takes some effort to deform a solid, as we have to break the bonds between the particles.

If we add energy, that is heat the system up, the particles will move around more vigorously. Eventually the particles will move around so much that they break the bonds mentioned above. We therefore change the phase from solid to liquid.[3] As a liquid, the particles will still "feel" the attractive forces of their neighbours, but they now move more freely among the other particles. The particles are still sticking together sufficiently to form a surface, but it is easy to deform a liquid.

Heating the system up further, the particles are now so energetic that they can whizz around freely. This gives rise to another phase transition, from liquid to gaseous. Now the particles move freely, unimpeded by the forces that try and keep them sticking together. Our example matter is now a gas.

Eventually particles move around so fast that the energy of the collisions is large enough to knock out electrons from the atoms that whizz around and bump into each other. We are left with a plasma, meaning that the atoms have lost one or more electrons in their outer layers due to the bumps, and are therefore electrically positive charged. This is shown in the panel furthest on the right of Fig. 5.2.

We can be more specific about how the particles are distributed on very small scales for the different phases of matter introduced above. As already mentioned, the typical size of an atom is of the order of 10^{-10} m, what used to be called an "angstrom". The distance between atoms in solids is therefore also of the order the size of the atoms, that is 10^{-10} m, since the particles are quite densely packed (that is, there is not much space between them).

---

[3]Changing from one state of mater to another, for example from a solid state to the liquid state, is also known as a "phase transition".

A useful quantity to compare, and also to get a feeling for the different states of matter, is the "mean free path". The mean free path is the average distance a particle can travel before bumping into another particle. It depends on the pressure and the temperature of the liquid or gas. The mean free path in water at room temperature is still of the order of angstroms, although the molecules are now moving around more freely, and there is nearly no space between the particles.

In everyday gases like air at room temperature and standard pressure the mean free path is roughly $0.1\,\mu m$, or $10^{-7}\,m$, that is about 1000 particle diameters. If we move to a more extreme environment where there are fewer particles per volume, particles can travel further before bumping into another particle. For example, in Earth's upper atmosphere air is very diluted and 100 km above ground, the mean free path is about 1 m.

In general, the physics on small scales gives rise to and determines the behaviour of matter on larger scales.

We can now clarify what we mean by "many particles" in the heading of this section. For example, how many particles are in a small cube measuring 1 cm on each side? To get a rough estimate on how many particles are in a cubic centimetre of matter, we can use our assumptions from above and take the particles to be tiny spheres with a radius of $10^{-10}\,m$, or a tenth of a nanometre. We then find that we can fit about 2×10^{23} into the cubic centimetre. That is indeed many particles, 10^{23} is a hundred thousand billion. This assumed that we pack the particles tightly together.

Our rough estimate is close to what we would find from a more detailed calculation. For example, under standard atmospheric conditions, there are about 3×10^{22} water molecules in a cubic centimetre of water. Later on we will see that the average matter density in the universe today corresponds roughly to one hydrogen atom per cubic metre.

5.1.3 Radiation

At first it might seem odd to list radiation as one of the matter constituents of the universe. After all we already saw in Sect. 3.2.1 that radiation is nothing but electromagnetic waves, for example visible light.

The reason we have to include radiation in our list of constituents of the universe is that electromagnetic waves carry energy. From everyday experience this is quite a familiar phenomenon. We mentioned in Sect. 3.2.1 that infrared radiation, consisting of waves with wavelengths slightly longer than visible light, will heat up a body. Another example is a microwave oven that uses electromagnetic waves with even longer wavelengths of about 1 mm to heat up liquids and food. But electromagnetic waves of all wavelengths carry energy.

We should add here that not only electromagnetic waves carry energy, this is common to all types of waves. That waves carry and transmit energy is also not outside our normal experience, which we illustrate with another example: a surface-wave on the surface of the ocean does not move water particles forward, unlike for example a stream of particles in a pipe. But it does carry energy, as any swimmer can attest to who gets lifted up by a wave.

But why do we highlight here that waves, and in particular electromagnetic radiation, carry energy? The distribution of energy in the universe is a key ingredient when we try to understand and model the evolution of the universe. For example, the rate of expansion of a region of the universe depends on the amount and the kind of energy present in this region. We will discuss and motivate these key concepts below in Chap. 6.

Another reason why we include radiation in our list of matter constituents of the universe, after discussing particles, is that electromagnetic waves can also behave like particles; these particles are called photons. We have used this fact already in previous chapters, using both the terms "radiation" and "photons". But why do we think waves can behave like particles? Even if they have properties in common like carrying energy this seems to be a rather strange assumption, and will be the topic of the next section, using radiation to begin.

5.1.4 Waves and Particles

A classic example that shows the particle nature of electromagnetic radiation is Einstein's explanation of the photoelectric effect, for which he won the Nobel prize in 1921. This effect is one of the physical phenomena scientists couldn't explain properly at the end of the nineteenth century. Another phenomenon was the shift of the perihelion of Mercury, that we encountered in Chap. 2.

So what is the photoelectric effect? If we shine ultra-violet light onto a metal surface during an experiment, we find that electrons are somehow extracted from the surface (here the colour of the light isn't important, as long as the wavelength is short). The experimental setup allows us to change the intensity, the amount of energy per second, and the wavelength (or frequency) of the light, and to measure the kinetic energy of the electrons that have been "liberated" from the surface.

The outcome of our experiment is that (1) the maximum kinetic energy of the electrons increases with the frequency of the radiation, and it does not depend on the intensity, (2) there is nearly no time lag between switching the light on and electrons getting freed from the surface, and (3) there is a minimum frequency below which no electrons are getting extracted. What does that mean and how can we interpret the results of this experiment? We know that the electrons need energy to be separated from the atoms in the surface of the metal (see the discussion of atoms at the beginning of this chapter).

Let us first try and explain the results assuming light is not a particle, and behaves *only* wave-like. In this case we would expect that the electrons should get extracted independent of the light frequency. Why should the process depend on the frequency? Also, if we wait long enough every free electron at the surface of the metal should eventually have caught enough energy from the light waves to break free, even if the frequency of the light is low. But we don't observe this. Also we can't explain the maximum kinetic energy of the electrons: if we increase the intensity, we would expect that also the electrons gain more energy.

On the other hand, let us assume now, as we did at the beginning of this section, that light is a particle called the photon, and the energy of the photon is proportional to its frequency. In this case things can be explained quite elegantly and naturally. The electrons have a maximum kinetic energy given by the photon energy (actually the photon energy minus the energy it takes to kick the electron out of the surface). There is no time lag because the electron doesn't accumulate energy from the light, it either gets kicked out, or it doesn't, depending only on the frequency of the light (that is the energy of the photon). Since there is a minimum amount of energy needed to kick the electrons out of the surface of the metal, there is also a minimum frequency below which the photons are not energetic enough to kick the electrons sufficiently, as the photon energy is proportional to the frequency. Note, that now the intensity is directly related to the number of photons reaching the surface (at a given frequency).

If we want we can consider the photons as being little wave packets, localised in space going in a single direction (we shouldn't think of them as little balls, but everybody else does ...), whereas in the wave picture the wave emanates from a light source in all directions.

That the energy of the photon is proportional to its frequency, or inversely proportional to its wavelength, is these days familiar even from our daily lives. For example, we use X-rays, electromagnetic radiation of very high frequency and hence very short wavelength, whose photons therefore also have high energy, for clinical purposes, because the high energy X-ray photons can pass through tissue and only get stopped by denser material like bones.

We have described above how electromagnetic waves can also behave like a particle. But that waves can behave like particles and particles like waves is not limited to radiation, it is common to all elementary particles and usually referred to as the "particle-wave duality": a particle can either behave like a classic billiard ball, using a rather naive picture, or it can behave like a wave, and hence show interference.

The classic observational evidence for this is the two slit experiment, which shows that elementary particles also behave like waves, in that they can form interference patterns. What we would "normally" regard as particles, would not show this behaviour. The experiment and its setup is detailed in Fig. 5.3. The slits are shown from the top, and next to them the pattern created on the screen opposite the slits, with the screen in frontal view.

The picture at the top of Fig. 5.3 shows the interference pattern created on a screen by waves passing through the slits; the waves could be surface waves on a pond, or light, for example. We already discussed interference above in

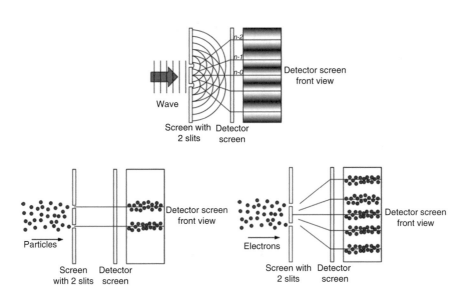

Fig. 5.3 A double slit experiment, showing that particles can also behave like waves. The slits are shown from the top, while the pattern created on the screen opposite the slits is shown in frontal view. The picture at the top shows the interference pattern created by waves passing through the slits, the waves could be surface waves on a pond, or light, for example. The picture on the lower left shows the pattern we would expect when billiard ball like particles are fired through the slits. The lower right picture shows what actually happens when particles such as electrons or even atoms are fired through the two slits, namely that we observe an interference pattern similar to the one in the top picture. *Image credit: plus.maths.org*

Sect. 3.3.4.2, in connection with the gravitational wave detector or telescope LIGO. Here we get an interference pattern on the screen, because two waves arriving at the point on the screen will have travelled a different distance from the left slit and from the right slit (with exemption of the point in the middle of the screen). We assume the waves from both slits come from the same coherent source, the waves therefore are in phase when they leave the slits. We get constructive interference on the screen when two wave crests reach the screen on top of one another, the amplitude of the waves adding up. We get destructive interference when a wave crest and a wave trough reach the screen together. If we use light for the waves in the experiment we get the distinctive interference pattern shown in the figure, a band like structure with dark and bright stripes, where we have destructive and constructive interference respectively.

The picture on the lower left shows the pattern we would expect when billiard ball like particles are fired at the slits: particles that behave like billiard ball cannot interfere, they either pass through the slit or not. The pattern on the screen would just be two "bright" bands opposite the slits (bright here means a higher density of particles).

The lower right picture shows what actually happens when particles such as electrons or even atoms are fired through the two slits: in this case, just as in the case of waves passing through the slits, we again observe an interference pattern, similar to the one in the top picture. We will get the same pattern even if we send individual electrons or atoms through the slits! The particles must interfere with themselves in order to generate the pattern, and hence they do behave like waves.

We can also associate a wavelength with the particles, known as the "de Broglie wavelength". Just as we associated energy with wavelength, or frequency, in the case of radiation and photons above, we find here that the more energy a particle carries, the shorter the corresponding wave length is.

How can we understand the particle-wave duality? We think it is fair to say that at the moment nobody has a real explanation of where this duality comes from. There are reasons of mathematical consistency of the underlying theory, quantum mechanics, that allow us to interpret objects as being waves *and* particles, but that doesn't really answer the question "how can a particle also be a wave?".

5.1.5 More Strange Quantum Behaviour

In the section above we introduced a key concept from quantum mechanics, namely the particle-wave duality. There is more quantum strangeness that we need to briefly discuss here, as it will be relevant later on in the book.

The reader might be familiar with Heisenberg's uncertainty principle, which states that the position and momentum, or speed, of a particle can't both be known to infinite precision. If we measure the position of a particle very precisely, then the speed is known only very roughly, and if the speed is known then the position is rather uncertain. This has nothing to do with limitations of our measuring devices. It is a fundamental principle, the product of the uncertainty in the position and in the speed is a very small, but non-zero number.[4]

Put another way, if we try and confine the particle in the speed dimension, it will be less confined in the position dimension, and vice versa. As a rather clunky analogy we can think of a blob of jelly between two plates of glass. If we squeeze the jelly under the glass in one direction (or dimension), it will be squeezed out in the other direction.

More important here is that the uncertainty principle can also be stated in terms of time and energy, instead of position and momentum; the two concepts position and momentum, and time and energy are in a sense similar. This means we can know the energy of a particle very precisely if we allow for a lot of time. Or, alternatively, if the time span allocated is very short, the uncertainty in the energy is very large. But as pointed out above, this has nothing to do with the measuring apparatus and its inaccuracies, it is a fundamental principle. We can also rephrase it, if the time scale of the fluctuation is very short, the energy fluctuation is very large. This in turn can then be interpreted to mean that for very short periods energy fluctuations, or particles, can fluctuate into existence and then vanish again. These fluctuations play a crucial role in the very early universe during inflation and are responsible for the formation of structure later on, as we will see in Chap. 9.

Although Heisenberg's uncertainty principle sounds bizarre and counter-intuitive, it has been experimentally verified, and it has so far passed all experimental tests. The uncertainty principle is these days a corner stone of physics, and cosmology. But there is more strangeness and weirdness to come

---

[4]The product of the uncertainty in the position and in the speed would have to be zero, if we could measure both to infinite precision. Experiments have shown this to be wrong.

as we shall see at the end of this chapter, which has not much to do with quantum mechanics.

5.2   Where Does Normal Matter Come From?

Let us now discuss where the normal matter, for example the atoms we are made of, the hydrogen and helium of stars, and the not so exotic elementary particles we see in high energy experiments, come from. In the following we focus mainly on particles like neutrons and protons, and atoms like hydrogen and helium, not only because they are familiar to the reader, but also because of their crucial roles in cosmology.

At the end of the inflationary period the universe was filled by a very hot and dense particle soup, including quarks, electrons and photons (see Chaps. 1 and 9). As the universe expanded it cooled down, and as the temperature fell the particles slowed down. Eventually the universe was cool enough and the particles slow enough for the primordial quark soup to "freeze out" and turn into protons and neutrons. This is the beginning of "nucleosynthesis", or the formation of atomic nuclei. As mentioned previously, the nucleus of hydrogen is a single proton.

While the universe is hot enough, protons and neutrons with the help of other particles transform into each other. However, when the universe cools, this becomes a one way reaction, neutrons decay into protons.[5]

When the temperature decreases further, the protons and neutrons bumping into each other form the nuclei of the light elements helium (two protons and up to two neutrons), lithium (three protons and up to four neutrons), and beryllium (four protons and up to four neutrons). This process lasts until roughly 20 min after the beginning. By that time the universe has expanded and cooled sufficiently for the colliding particles and nuclei to be too slow and therefore not having enough energy to overcome the repellent forces of the positively charged protons.

The process also stops because a free neutron, one that is not bound yet into a nucleus, has only a finite lifetime unlike for example the proton and the electron. The neutron has a half-life of roughly 10 min, which means that after 10 min out of 100 free neutrons only 50 are left, the rest decayed into a proton,

---

[5]The reactions are, for example, neutron ↔ proton+electron+neutrino, while it is hot in both directions, when it is cooler only from left to right.

an electron and a type of neutrino.[6] Once the universe is sufficiently cool, if free neutrons are not bound in a nucleus, they decay. Luckily those neutrons bound together with other neutrons and protons in nuclei, forming helium, lithium, beryllium, survive, but the rest of the neutrons decay into protons and other particles. The left over protons are hydrogen nuclei.[7]

No elements heavier than beryllium are formed in the early universe, as the nuclei of heavier elements aren't stable enough to survive the hot and highly energetic environment in the early universe. By the time it is cool enough for them to survive, the particles are too slow to overcome the repellent forces of the protons.

To distinguish this early process from the formation of heavier elements later on in the history of the universe in stars and supernovae, it is also often referred as "primordial nucleosynthesis" or more sloppily "Big Bang nucleosynthesis". At the time, the universe was still hot enough to ionise the light elements, that is separate the electrons from the nuclei, and so all normal matter was still in plasma form. The universe expanded and cooled down, and only when the universe as about 380,000 years old was the temperature low enough for the last hydrogen nuclei to combine with the remaining electrons to form neutral hydrogen. Since before this time the photons were still "coupled" to the free electrons—the photons and the electrons interacted with each other—this point in the history of the universe, when the electrons and photons stopped interacting, is also known as "decoupling". It is also known as "recombination", since after decoupling the hydrogen nuclei can combine with the free electrons. We will return to these topics in Sect. 8.2.3.

The heavier elements, like carbon, nitrogen, oxygen, iron, and the rest of the periodic table, were not formed in the early universe. They are produced later on in supernova explosions of stars. That is, the stuff crucial for planets, human beings and life in general was only formed after the first stars had formed and died.

Today hydrogen is still the most abundant element in the universe, followed by helium. Roughly 75% of the baryonic matter is in the form of hydrogen, and 24% in form of helium. The rest totals just about 1% of the normal matter, which here we could consider mere "impurities", but is nevertheless essential for planets, life, and the universe as we know it.

---

[6]The neutron decays into a proton, electron and an anti-neutrino. The anti-neutrino is the antiparticle of the neutrino.

[7]The universe would be very different if the bound neutrons would also decay, as only protons and electrons would be left to form atoms from, and only hydrogen atoms would exist.

5.3   How to Categorise Different Types of Matter

Before we move on to "non-normal" types of matter, namely dark matter and dark energy, in the next sections we briefly discuss how cosmologists often classify and group together different types of matter. The physical properties used to distinguish between these groups all depend on the behaviour of the constituent particles of the type of matter under consideration: (1) whether the particles interact with electromagnetic radiation, (2) whether they exert pressure and if so what kind of pressure, and (3) what dominates their energy.

1. *Luminous versus dark*: the particles from which normal baryonic matter is made of, for example protons, neutrons, and electrons, can interact with electromagnetic radiation, such as light. We therefore literally can see normal matter, that is an object emits or reflects light, and we "detect" it with our eyes or some more technologically involved means. Normal matter therefore counts as "luminous".

   However there are also particles that do not interact with electromagnetic radiation. These particles and the matter they make up are therefore referred to as "dark".

   What about our everyday definition of "dark" applied to an object, that is not to emit or reflect radiation? If normal, baryonic matter is extremely cold, at zero absolute temperature, or 0 K, it will not emit radiation. However, if the temperature is above 0 K it will inevitably emit thermal radiation, typically in the infrared part of the electromagnetic spectrum. It is therefore nearly impossible for baryonic matter to be completely dark.[8]

2. *Pressure*: using our simplistic particle model again, we can describe pressure as the force exerted by particles bouncing off the surface of an object. To be more precise, pressure is the aggregate force exerted by very many particles bouncing off a surface, per surface area. For example, the water molecules bouncing off the walls of a pressure cooker can give rise to considerable pressure.

   However, in order to exert pressure the particles have to interact with the surface of the object. If they can simply pass through the surface, they wouldn't give a rise to pressure. Also, we would expect the pressure force to depend on the strength of the impact or bounce off the surface, the faster a particle hits the surface the larger the force and hence the pressure will be. As

---

[8]The Cosmic Microwave Background has today a temperature of 2.73 K and is present or "shines" everywhere in the universe, keeping normal matter above absolute zero even in the absence of stars.

discussed earlier in this chapter, the speed of particles usually increases with temperature, and therefore we expect that also the pressure to increase with temperature. This in agreement with our normal experience, for example the pressure in our pressure cooker will rise if we increase the temperature.

This also means that if particles move very slowly, the pressure they give rise to will be very small. Therefore very heavy particles, moving around very slowly, can be modelled as a gas with vanishing or zero pressure.

As well as non-zero and zero pressure, for cosmologists there is a third option, namely negative pressure. This has no equivalent in "normal" physics and shouldn't be mixed up with negative *relative* pressure, which is the difference between two positive pressure values, for example at the suction tube of a pump. Negative pressure, in the cosmological sense, can give rise to an accelerated expansion of the universe. Weird forms of energy and particular fields have this property as discussed below and in the next chapter.

3. *Relativistic versus non-relativistic*: the overall energy of a particle has a contribution from its kinetic energy, and from its rest-mass energy (in the simplest case). Kinetic energy is the energy of the particle due to its motion, rest-mass is the mass an observer at rest relative to the particle will measure, our everyday concept of mass. However, Einstein pointed out that mass can also be regarded as energy, rest-mass energy is this mass converted into energy (see the next chapter).

An object at rest will only have rest-mass energy, as we would expect from the name, and this energy is always the same for the object. But the faster the object moves, the larger its kinetic energy will be and therefore also the larger the contribution of the kinetic energy to the total energy. We can therefore distinguish between particles whose energy is dominated by the rest-mass, called "non-relativistic" particles, and particles whose overall energy is dominated by the kinetic energy, named "relativistic" particles.

A heavy particles which has a large rest-mass, will usually also move around fairly slowly and will therefore have a rather small kinetic energy. It is therefore a typical non-relativistic particle. A typical relativistic particle has a small mass and moves about very fast, therefore their overall energy will be dominated by their kinetic energy. Fast in this context means close to the speed of light. Only particles with zero rest-mass can travel with exactly the speed of light.

We are used to particles that have rest-mass, but rest-mass is not necessary for "objects" to behave like particles. If they have zero rest-mass, travelling at the speed of light, they can still have kinetic energy. This is indeed the case for electromagnetic waves, which travel at the speed of light, have zero rest-mass and can also be viewed as particles, which as pointed out in Sect. 5.1.3 are called photons. The energy of the photons depends on the frequency (or wavelength) of the radiation. The higher the frequency (shorter the wavelength), the higher the energy of the photon. Because photons are the particle associated with electromagnetic waves, they travel at the same speed, namely the speed of light.

5.4   Exotic Stuff: Dark Matter

The matter content of the universe we discussed so far, baryonic matter, or the stuff we are made of, and radiation is familiar from everyday life. The types of matter we discuss in this section and the next, dark matter and dark energy, play a lesser role in everyday life, but they are crucial if we want to understand the evolution and the dynamics of the universe. In fact they are so essential to the cosmological standard model, that specific examples of dark matter and dark energy have become synonymous for the model, as we shall see below.

5.4.1 Observational Evidence for Dark Matter

Dark matter is an exotic form of matter that has been postulated by theorists to reconcile their theories with the observational data, but which has, so far, not been observed directly. Currently it can be studied only through its gravitational properties, and ever since its first indirect detection it has been a curiosity. Both dark matter and ordinary matter are sources of gravitation and interact through gravity. Strangely dark matter does not interact in any other way that we know of or can detect, with the possible exemption of the weak force. Ordinary matter on the other hand can carry electrical charges and generate magnetic fields, and behaves as we expect matter to behave.

In 1933 Zwicky,[9] measured the velocities of eight galaxies in the Coma cluster of galaxies[10] and he used the results to calculate the mass of the cluster, applying the method described below. The value he calculated came out about 400 times the mass obtained using the mass of the luminous matter, and subsequent studies confirmed Zwicky's result albeit correcting to "only" about 20 times the luminous matter. Zwicky called the extra matter in the Coma

---

[9]Fritz Zwicky (1898–1974), a Swiss physicist and astronomer, worked mainly at Caltech, introduced "dark matter" to cosmology.
[10]The Coma cluster is a large cluster of galaxies, located about 100 Mpc or 300 million lightyears from Earth, containing roughly 1000 galaxies. It is similar, but larger, to the local group of galaxies, see Fig. 4.10.

cluster "*dunkle (kalte) Materie*" which can be translated as "dark (cold) matter". This was the first detection of dark matter, however it remained a curiosity and didn't attract much attention in the research community.

Dark matter was finally accepted by the astronomical community when Vera Rubin[11] in the 1970s published her observations of galaxy rotation curves. Rubin used the same method as Zwicky, but measured the rotation velocities of stars in galaxies, whereas Zwicky used the velocities of galaxies in a galaxy cluster. Dark matter is now an accepted part of the total energy and matter budget in the universe.

How do astronomers calculate the mass in a galaxy from the velocities of the stars of the galaxy? We have to anticipate our discussion of gravity in Sect. 6.4.1 of the next chapter, and remind the reader of Newton's inverse square law, which gives rise to a simple distance-rotation speed relation: the rotation speed of a small mass orbiting a large mass depends on the distance from the centre of the large mass. The further the small mass is located from the central mass, the slower it will move around the central mass, the velocity will fall like the square of the distance, hence the name "inverse-square law". For example if the small mass is twice as far from the central mass it will move a quarter as fast. This behaviour is also known as "Keplerian motion".

In this context the small mass is a star and the large mass is the bulge or central part of a spiral galaxy, consisting of millions of bright, luminous stars. From observations, that is simply counting stars, we know that most of the luminous matter in a spiral galaxy is concentrated in the central part or bulge. We would then expect, according to the inverse-square law, that the further the star is located from the centre, the slower it will move around the central mass.

This is illustrated in Fig. 5.4, which shows the rotation speed of a star in a typical spiral galaxy. The origin of the graph is at the centre of the galaxy. Now if we plot the observed speed of the star at various distances from the centre then we expect to get a graph like the lower (dotted) one in Fig. 5.4.

However, the observed rotation curves of disc galaxies are of type Fig. 5.4 with the curve flat well beyond the limits of the luminous matter. In order to explain the observations, Zwicky suggested that there is an unseen, or dark, matter component providing more mass in addition to the luminous matter. Rubin's observations then showed that the dark matter encompasses the galaxies in a roughly spherical cloud, called a "halo", as shown in Fig. 5.5.

---

[11]Vera Rubin (1928–2016), US astronomer, her work on spiral galaxy rotation curve established dark matter.

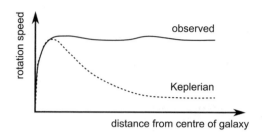

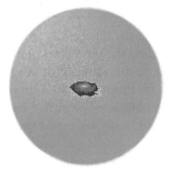

Fig. 5.4 A typical galaxy rotation curve: on the vertical axis is the rotation speed of a star orbiting the centre of the galaxy, which is taken to be at the origin of the graph (where the axes meet), on the horizontal axis is the distance of the star from the centre. The top curve is the actual measured or observed rotation curve, the dotted lower, steeply falling off one is the rotation speed expected for Keplerian motion

Fig. 5.5 A typical dark matter halo: the galaxy in the centre is surrounded by a roughly spherical cloud of dark matter, the "halo" of the galaxy. All galaxies, whether they are spiral galaxies or elliptical in shape, are surrounded by a dark matter halo. The diameter of the halo is at least 5 times larger than the diameter of the galaxy

This halo extends to distances roughly five times further than the edge of the visible matter of a galaxy. The dark matter halo is always roughly spherical in shape, irrespective of the type of galaxy it surrounds. Our own galaxy the Milky Way is also surrounded by a halo of dark matter. In Fig. 5.4 the end of the rotation curves corresponds to the edge of the luminous matter, that is the outer rim of the galaxy. In the case of the Milky Way the visible matter extends to about 50 thousand lightyears, but the dark matter halo has a radius of very roughly 250 thousand lightyears.

Although we discussed only spiral galaxies, in fact all galaxies are made up of luminous, or baryonic, matter and dark matter with there being about 20 times more dark matter than luminous matter. The dark matter is needed to explain the observations, it provides the additional gravitational mass or "glue" that hold the galaxies, and galaxy clusters, together.

Large scale structure, the distribution of galaxies and clusters of galaxies on the largest scales of hundreds of Megaparsecs and more, provides another piece of observational evidence for the existence of dark matter. Cosmologists began in the 1980s to use computer simulations to study the formation of the large scale structure. These simulations showed that to get a similar distribution of galaxies to what we observe today, see for example Fig. 4.9, we need dark matter, as the luminous material doesn't provide enough gravitational attraction. We will explain in some more detail in the following chapters the essential role dark matter plays in the theory of galaxy and structure formation in the universe.

As the nature of the dark matter is still a mystery, we can only assert some very general properties of the dark matter with the help of the definitions and broad categories introduced in Sect. 5.3. However we will see in the following section, there is one property all dark matter candidates have in common—they are dark.

5.4.2   Dark Matter Candidates: WIMPs, MACHOs, and Black Holes

Besides being dark, what other properties do we know that dark matter must possess? It has to have mass in order to interact gravitationally, and there has to be a lot of it, we saw above that there is 20 times more non-luminous matter in a galaxy than luminous matter. But this is nearly all the information we have. It is frustrating that we do not have a proper understanding of dark matter at present, but on the other hand that leaves open a large search area to try and find a model.

We can now discuss the candidates or groups of candidates that satisfy these very broad conditions, starting with the most obvious one: could dark matter be just normal baryonic matter that doesn't emit electromagnetic radiation? This would be a very simple solution, and in line with Zwicky's original hypothesis of cold, dark matter (it has to be cold, or it would emit thermal radiation). Hence the baryonic matter could be in the form of clouds of cold gas, but the clouds have to be cold, or they would emit radiation.

Alternatively, if the clouds have collapsed under their own gravity, they might have formed objects similar to the planets of the solar system. These "failed stars" are known as *Massive Astrophysical Compact Halo Objects*, abbreviated *MACHOs*, planet sized objects, with masses from around one thousandth of the mass of the Sun, to masses around eight hundredths the mass of the Sun. MACHOs don't emit light and cannot be seen directly (since they are failed stars). There is however a neat, indirect method of detecting them which is to see the effect that occurs when a MACHO passes in front of a bright light source. The effect of the unseen MACHO is to focus the light from the source and so cause a temporary brightening of the light. While the MACHO itself cannot be seen there is indirect evidence for its existence and MACHOs have been detected in this way.

Unfortunately there are two problems: first, astronomers looked very carefully into baryonic dark matter, but haven't found observational evidence for sufficient amounts of cold gas and failed stars. But even if we would assume that there is more "hidden" baryonic matter that hasn't been observed yet, there is another more fundamental problem. The amount of baryonic matter produced during primordial nucleosynthesis, described above in Sect. 5.2, is too small. The maximum amount of dark baryonic matter can only make up about 5% of the total dark matter, to fit into all the theoretical and observational constraints.

As an alternative to the assumption that normal matter forms compact objects like MACHOs we could assume that the dark matter consists of a population of small black holes. An idea that can be taken even further by postulating that the black holes are primordial, in other words they were produced in the first fractions of a second after the beginning, well before nucleosynthesis. This is therefore a viable possibility, but again, the observational evidence for sufficient amounts of these "primordial black holes" is missing (they could be observed in a similar indirect way as MACHOs). We will discuss these objects again in the next chapter.

If MACHOs and black holes are not the solution, we have to look for non-baryonic dark matter, exotic stuff. One of the most popular dark matter candidates are called WIMPs, or Weakly Interacting Massive Particles. A typical WIMP particle has a mass in the range of about ten to a few thousand of proton masses and would have been formed shortly after inflation, and they now need to remain in sufficient numbers to be the dark matter. There are many hypothetical elementary particle candidates for this non-baryonic dark matter. Attempts have been made to create WIMPs in accelerators or to detect them directly as they pass through the laboratory. So far without success, which does not mean that WIMPS are not the solution, but they are beginning to become slightly less popular.

Exotic particle dark matter can be further differentiated according to the mass of the WIMP and how fast it moves, as discussed in Sect. 5.3. Heavy particles, moving around sluggishly are referred to as non-relativistic, whereas particles with small mass moving about very fast are referred to as relativistic. Since hot temperature are associated with large particle speeds, and cold temperatures are associated with small speeds, the particle dark matter is also referred to as warm or cold. Although not excluded, numerous studies using warm dark matter models have failed to provide a better model than cold dark matter.

Most of the candidates for dark matter models fall into one or the other of the two classes, MACHOs or WIMPS, discussed above.[12] While each of these models has its attractions, and at the moment there is still no fully agreed upon explanation or model for the dark matter, although it is fair to say that most cosmologists tend towards a WIMP like model.

5.5 Weird Stuff: Dark Energy

There is one more exotic component in the total energy budget of the universe to compete in weirdness with *dark matter*, and that is *dark energy*. All we know about it is that it is a form of energy, and it causes the universe to expand at an accelerated rate. This is very unusual, as in most sensible models of a universe filled with radiation, normal matter and dark matter, the universe is expanding but its rate of expansion will slow down. Dark energy also has negative pressure, which makes it "weird", as normal matter can only maintain positive pressure. It is "dark" as it doesn't interact with normal matter, it might not even interact gravitationally.

5.5.1 What Is the Observational Evidence for Dark Energy?

Dark energy was first observationally confirmed in the 1990s by a team studying the expansion of the universe, using the method discussed in Sect. 4.3 to measure distances, which makes use of supernovae of type Ia (SN Ia) as standard candles. The team found that the supernovae were at larger distances than what was expected if the universe was expanding at a decreasing rate. We will discuss this in detail in Sect. 8.2.1. It took a few years for this surprising result to be fully accepted. However, independent studies of for example the

---

[12]The acronyms "WIMP" and "MACHO" are these days firmly established in cosmology, reflecting the rather special sense of humour cosmologists possess.

Cosmic Microwave Background and the large scale structure support the need for an unidentified extra energy contribution in the universe.

This need arises in the following way. Measurements of the distribution of small temperature fluctuations in the Cosmic Microwave Background indicate that the universe is geometrically flat. But this requires an additional contribution to the energy budget of the universe, and today ordinary matter, or baryonic matter, and Cold Dark Matter together make up roughly 30% of the required density leaving roughly 70% unaccounted for, unless we include the extra energy density which is also demanded to explain the acceleration of the universe detected by the supernovae observations. The large percentage shouldn't be a worry as the energy is spread uniformly across the whole universe. As with dark matter there are few clues as to what this weird stuff actually is. In the next sections we will describe two popular candidates.

5.5.2 Candidates for Dark Energy

Before we discuss dark energy candidates it is useful to introduce another concept from theoretical physics that might be new to some readers, namely the notion of "field" as used in cosmology, and indeed physics in general. A field is another way to describe matter, in addition to modelling it as particles and as waves. Although it is often used for particles that are related to mediate or carry forces, like the photon in the case of electromagnetic field, also "normal" particles can be described as a field, for example we could talk about the electron field. We will illustrate the basic concept of a "field" with a simple example, and discuss fields in more detail in the following chapter. Consider the temperature of the air in a room. At each moment in time and at each point in space the air will have a temperature, and we can associate a "temperature field" with these measurements. The temperature field will change in time, if we for example switch the heating on, and it will also be different in space, that is different points might have different temperatures, for example depending on how close they are to the heater.

Besides the temperature in the room, as another example, we can also regard the air density in the room as a field. The main point is, the field has a certain value at each point in time and space. But let us return now to our discussion of models of dark energy, beginning with the simplest and, arguably, most popular model: the "cosmological constant".

In some sense, the cosmological constant is the simplest type of field: it is a field that is constant everywhere in the universe and also constant in time, although one might consider this stretching the field concept a bit. The "cosmological constant", often simply referred to as "Lambda" or Λ after the Greek letter Einstein used for it in his equations. When Einstein first proposed his equations for gravity he included an extra term, the cosmological constant. At the time it turned out that his and other models of the period did not need this constant. Recently, however, cosmologists used it to explain the accelerated expansion of the universe today, and therefore as a model of Dark Energy. It is usually called "Lambda" after the Greek letter (Λ) used to denote it. For this reason the cosmological model which includes Cold Dark Matter and uses a cosmological constant for the Dark Energy is called the ΛCDM model. At the moment, the ΛCDM model is considered by many cosmologists as part of the "standard model of cosmology". The physical effect the cosmological constant has is to give rise to a negative pressure, which drives the accelerated expansion of the universe.

This looks very good except that we appear to have replaced one unknown quantity, dark energy, with another, the cosmological constant. However, physicists working in Quantum Field Theory have produced models which attribute an energy to the vacuum—the energy of the quantum vacuum— but unfortunately the value is way too large for the cosmological constant.[13] Despite these problems the cosmological constant, (Λ), is the simplest model to explain the late time cosmic acceleration. It is a critical element of the current standard model in cosmology and as new observational evidence becomes available it continues to provide a good fit to the data.

Instead of adding or using a field which is constant throughout space and time, that is the cosmological constant, we can also introduce a field which varies through time and space. This route is taken by another approach to model dark energy, namely to introduce a dynamical field, which has been named "quintessence", and this field drives the acceleration.

Several questions arise: (1) Can a field with the required properties be constructed? (2) Does the field make sense in physics? (3) Once the field is included does the cosmological model still fit? (4) Can the quintessence field be chosen so that the accelerated expansion only starts at a late enough time to match observations? Quintessence fields with these properties can be constructed. There is a class of fields, called scalar fields, which have the property that if they change very slowly in time, they can give rise to a negative pressure, just what we need to model dark energy.

---

[13]The value for the cosmological constant found taking some quantum field theory calculation at face value is roughly 120 orders of magnitude too large, that's "12" followed by 118 zeroes, a very large number even by cosmology's standards!

However, the problem is that there is no other evidence yet for the existence of the quintessence field besides the late time acceleration of the expansion of the universe, and again it looks like we have replaced one unknown quantity, dark energy, with another, this time the quintessence field. Quintessence therefore remains a possibility and leaves open the possibility of an exotic model for dark energy, but it is more complicated than the cosmological constant.

In the next chapter we will explain some of the concepts introduced above in more detail, or at least we shall try. In particular we have to discuss what forces are at play in the universe that govern its evolution, and what do we mean by its expansion.

6

What Are the Forces That Shape the Universe?

In the previous chapter we discussed the energy content of the universe, the different types of matter that we see, or actually don't see. So far we have only briefly mentioned what forces come into play and what the nature of these forces is. We also haven't discussed how the matter constituents interact with each other and what forces govern these interactions, from the smallest to the largest distance scales. We shall do this in the following, beginning with a discussion of what forces there are, and then move on to what a force actually is and discussing finally how these forces shape the evolution of the universe.

6.1 How Many Forces Are There?

Luckily for physicists and cosmologists, the forces that govern the motion of a microbe on earth, determine the workings of a fridge, and control the orbits of the planets in the solar system, also govern the evolution of the universe on the largest scales: the same forces at play in a laboratory on earth are also responsible for the formation and evolution of stars, galaxies, clusters of galaxies and beyond.

At present physicist are aware of four forces, acting on different scales and with very different strengths. Four is a surprisingly small number which nevertheless gives rise to all the phenomena we observe in the universe. The number of forces cosmologists have to consider on very large scales is even smaller, namely just one, and that is gravity.

© Springer Nature Switzerland AG 2019
K. A. Malik, D. R. Matravers, *How Cosmologists Explain
the Universe to Friends and Family*, Astronomers' Universe,
https://doi.org/10.1007/978-3-030-32734-7_6

Table 6.1 The four forces, their strengths and ranges

Force	Strength	Range	Charge	Carrier
Strong nuclear force	1	10^{-15}	3 "colour" charges	Gluons
Weak nuclear force	10^{-5}	10^{-18}	2 weak isospin charges	W and Z bosons
Electromagnetic force	1/137	Infinite	Electric charge	Photon
Gravity	10^{-38}	Infinite	Mass	–

The strength of the forces are relative to the strong force, the range is in metres. We should point out that the "colour" charge has nothing to do with everyday colour. Also, to avoid complaints from particle physicists, the weak force and the electromagnetic force are not completely independent

But let's have a brief closer look at the other three forces, the strong force, the weak force, and the electromagnetic force, before we focus on gravity. We will be summarising some of the details of the forces, their range, their strength relative to the strong force, what charges are involved, and finally what particle or mechanism is involved in transmitting or communicating the force in Table 6.1. The charges in the table determine whether a particle or an object experience the force associated with this charge. Hence, for example, all massive objects are subject to gravity, but only objects with electromagnetic charge will experience the electromagnetic force.

From the table we see that only the "electromagnetic force" and "gravity" have an effect on large distance scales. The "strong force" and the "weak force" are at play on the shortest distance scales relevant inside the nucleus of an atom, the strong and the weak force are therefore also referred to as nuclear forces. As discussed in Sect. 5.1.1, a hydrogen atom is roughly 10^{-10} m or a tenth of a nanometre in diameter and a proton, the nucleus of the hydrogen atom, measures about 10^{-15} m across. The strong force is, unsurprisingly, the strongest force of the four, but acts only on very short distance scales, across a proton or neutron, gluing the constituents of these particles, the "quarks", together. The quarks carry a charge, called "colour", and different colours attract or repel each other. However, the quarks that make up particles like the proton and neutron have charges that sum up to be colour neutral; this concept is similar to the more conventional electromagnetic charge neutrality discussed below. We should point out that the "colour" charge has nothing to do with everyday colour.[1]

On the smallest scales, a little bit below 10^{-15} m, the strong force glues the constituents of the proton and the neutron, the quarks, together. As discussed

---

[1]This is another example of physicists' sometimes quirky sense of humour. The three different colour charges combine to colour neutral, in a similar way as the three primary colours combine to white.

in the previous chapter, protons and neutrons consist of three quarks each. But there are many other compound particles, consisting of two, three or more quarks, all held together by the strong force. On slightly larger scales, as far as it reaches, roughly 10^{-15} m, the strong force is also responsible for keeping protons and neutrons together in the nucleus of an atom.

The weak force is much weaker in strength than the strong force. Its charge is called "isospin" and the weak force has an even shorter range than the strong force, namely about 10^{-18} m. It governs, for example, the decay of the neutron into a proton. Besides nucleosynthesis, as discussed in Sect. 5.2, it plays no direct role in cosmology. The only exception here are neutrinos, as they only interact through the weak force, as mentioned in Sect. 3.3.4.1. This means that neutrinos can travel large distances without interacting with the matter between the source and an observer on earth, but it unfortunately also means they are very difficult to detect once they have reached earth. In both instances, the weakness of their interaction is to blame.

Until recently, neutrinos were assumed to be massless, like the photon. In 2001 experiments in the Sudbury Neutrino Observatory convinced physicists that neutrinos are not massless, but have a very small but non-zero rest-mass, the size of which is however still not determined and a topic of active research. Neutrinos do play a role in the overall energy budget of the universe at very early times, despite their tiny mass because there are so many of them, roughly 340 million per cubic metre today. Due to their small rest mass neutrinos behave similar to the photons or radiation. They can therefore make a small, but measurable contribution to the overall evolution of the universe on large scales.

Before we irritate too many particle physicists, we should point out that the weak force is not completely independent of the electromagnetic force. In the Standard Model of particle physics, the weak force and the electromagnetic forces are united as the electro-weak force. This becomes experimentally manifest at very high energies or temperatures. Indeed at even higher energies the Standard Model allows to describe the strong and the electro-weak force as different aspects of one underlying unified force. But let us return now to the discussion of the four forces in their non-unified versions.

We don't need the weak and strong force to understand the universe on large scales, however they are essential to understand nucleosynthesis, the evolution of stars, and other "messy stuff". The main reason the strong and the weak force are not really relevant for cosmology after nucleosynthesis is that their range is so small. After all, we need to consider length scales of lightyears or roughly 10^{16} m, or even billions of lightyears or Gigalightyears, that is 10^{25} m, and therefore forces acting on scales of up to 10^{-15} m do not really matter, with exception to the very beginning of the universe as will be discussed later on in Chap. 9.

The other two forces, the electromagnetic force and gravity, have much larger ranges, and in principle their range is infinite. Let us have a closer look at the electromagnetic force next, it is probably also more familiar to the reader than the nuclear forces, we will postpone our discussion of gravity to later.

6.2   The Electromagnetic Force

We are all familiar with applications or effects of electromagnetism, for example the electric current that provides light in a lamp, the magnetism that causes magnets to attract pieces of iron and some other metals and the imperceptible signals which drive our televisions. The physical theory which explains all these effects and others that we will come to below and later is electromagnetism.

It is quite remarkable how we have grown familiar with the concept of the electromagnetic force, which unifies the electric and the magnetic forces. We usually don't think about the fact that it was not until the eighteenth and nineteenth centuries that most physicists stopped considering the magnetic and electric forces to be completely separate phenomena.

The effect of electric charges, the electric force, is familiar from every day life. It manifests itself for example when crossing a woollen carpet in rubber soled shoes and then touching a metal door knob: we might get a little shock, even producing a small spark when touching the metal. This is an example of static electricity. Another example is combing dry hair with a plastic comb. If we keep combing for a while, the hairs will start separating from each other, being repelled by the static electricity, or to be more precise, by the electric force.

How can we describe the electric force? Performing some simple experiments, not much more intricate than the examples mentioned above, we would conclude that there are two types of charge, positive and negative ones. For example the electron carries one negative charge, whereas the proton carries one positive charge. If an object carries the same number of positive and negative charges, the object is charge neutral. Two objects will experience the electric force depending on the charge on each object, and the distance between the two objects. The force experienced by the objects is proportional to the excess charge on each object, and proportional to the inverse of the square of the distance between the objects. The objects will repel each other, if the charges have the same sign that is both are positive or both negative, and the objects will be attracted towards each other if the charges have opposite signs, as shown in Fig. 6.1. If we take a charged object, either positive or negative, and put it next to a charge neutral object, neither will experience an electric force.

For example, the hydrogen atom consists of a single proton as nucleus, and a single electron, see Fig. 5.1 in the previous chapter. The electric force attracts the electron and the proton to each other, this is what keeps or binds the two particles together. But the hydrogen atom overall is charge neutral. Hence, although there are really a lot of charges in a litre of hydrogen at room temperature, since there is the same number of positive as negative charges, the hydrogen gas wouldn't "notice" a charge put next to it.

The electric force keeps electrons bound to the nucleus of an atom, as shown in the cases of the two lightest and also most abundant elements hydrogen and helium in Fig. 5.1, and this is also the case for all other, heavier elements in the periodic table. But the electric force also binds atoms into larger molecules at microscopic scales, and binds molecules and atoms together into macroscopic objects. This is done through sharing of electrons in the outer "layers" of the atoms. It plays therefore an essential role in everyday life.

Before turning to magnetic phenomena, we should briefly discuss how the electric force gets transmitted between two charged particles, as in Fig. 6.1, for example. To answer this question we need to anticipate some of the material discussed in more detail in Sect. 6.3 below. Every electrically charged particle is surrounded by an electric field, and it is this field that communicates the electric force between charged particles, as a result the particles attract or repel each other. The electric field is sketched in Fig. 6.2 for the case of a positive and negative charged particle. As already mentioned, the two particles attract each other, and this attraction is mediated through the electric fields of each particle. The field shown in the figure is the superposition of the two electric fields of each charged particle.

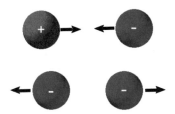

Fig. 6.1 Illustrating the electrostatic force: in the top row two spheres with opposite charge attracting each other. Bottom row, two spheres with the same charge, here both negative, repelling each other

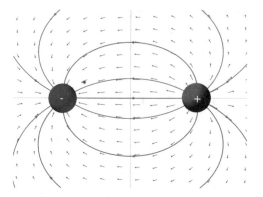

Fig. 6.2 Illustrating the electric field generated by a positive and a negative charged particle. The field strength can be seen from the density of the field lines (in red). The field is strongest close to the particle and falls off with the inverse of the distance. The electric field also has a direction, from the positive charge to the negative (indicated by the arrows)

This is due to a remarkable property of the electromagnetic force and therefore of charged particles, namely that we can superimpose their individual electric fields. The net electric field of, for example, a proton and an electron cancels out in between the charges in such a way that they attract each other. Then a single negative charge will attract a single positive charge with a force equal to a constant times the product of the charges divided by their distance apart and vice versa. Further away from the two opposite charged particles there will be no effect on the environment, the effect of the charges is "screened off", the electric fields of the two opposite charges cancel each other. Two equal charges either positive or negative will repel each other with a force again equal to the product of the charges divided by their distance apart. All these effects are mediated through the electric field. The field is strongest close to the particles and falls off with the inverse of the distance from the centre of each particle, as discussed above. We will discuss the electric field and fields in general further in Sect. 6.3 below.

Let us now have a closer look at the magnetic force, the force that is experienced by magnetic objects, and how can we describe it. "Magnetic objects" or magnets are quite common in everyday life, examples might be the ubiquitous fridge magnets or the bar magnets sometimes found in a workshop (very useful to pick up difficult to reach screws and similar small metal objects).

Performing some simple experiments with bar magnets, we would conclude that there seem to be two types of "charge", as in the electric case above, with opposite magnetic charges at each end of the bar magnet. If two magnets are brought together then the poles with opposite charges attract one another and those with the same magnetic charges repel each other. Using the standard terminology, one end is referred to as "magnetic south pole", the other as the "magnetic north pole", see Fig. 6.3. North and south poles attract each other, whereas poles of the same "charge", repel each other.

However, there are no magnetic "south" and "north" charges, unlike in the electric case, where we do have positive and negative electric charges! Contrary to the electric charges which can occur in isolation, magnetic poles cannot and always occur in pairs. If we cut the bar magnet into half, we don't end up with the "north" and the "south" part, instead we end up with two bar magnets, half the length of the original magnet, each with a "south" and a "north" pole, see Fig. 6.4.

How can we reconcile our experimental results, that there seem to be magnetic charges, with the fact that they only come in pairs? This brings us also, maybe surprisingly, to the question of whether and how the electric and the magnetic force are related. The answer to this question is simple: moving electric charges generate a magnetic field. This by no means obvious experimental fact was discovered in the nineteenth century.

An electric current is simply some moving charged particles, for example negatively charged electrons moving through a copper wire. However, any

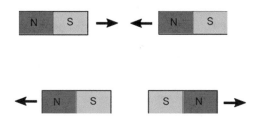

Fig. 6.3 Illustrating the magnetic force using bar magnets. In the top row, the north and the south poles attract each other. Bottom row, two bar magnets facing each other with the same pole, here both south, repel each other

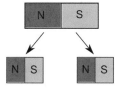

Fig. 6.4 Illustrating what happens if we cut a bar magnet in half: we get two bar magnets, each with a north and a south pole and each with half the strength of the original, and not the north and the south part of the original magnet

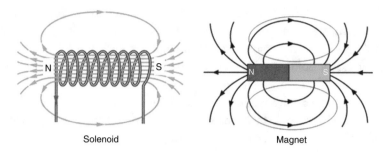

Soleniod   Magnet

Fig. 6.5 Illustrating the similarities between a bar magnet (on the right), and a coil with an electric current flowing through it (on the left). Both magnet and coil generate the same field line pattern. If we cover both with a sheet of paper, and spread some iron filings across the sheet, we would see a very similar pattern formed by the iron filings tracing the magnetic field lines on both magnets. *Image credits: OnlinePhys*

moving charged particle can be referred to as current. If a current of electricity flows through a wire near a compass, the compass needle will be deflected, because the current, that is the moving charges, generate a magnetic field.

If we wind the wire into a coil and have a current flow through it, the magnetic effects of the individual wires will add up and the coil will behave like a bar magnet, see Fig. 6.5, the coil is also referred to as a "solenoid". The coil also has a north and a south pole and will attract or repel other magnets accordingly. We can now introduce the concept of "magnetic field" in a similar way as we did above with the electric field to discuss how the magnetic force is mediated. Below we will discuss fields in more detail, but preempting this discussion we note here that the magnetic field line pattern from the coil and bar magnet are very similar, Fig. 6.5.

We use this to explain the behaviour we observe when cutting a bar magnet in half, and we get two bar magnets, with north and a south pole, instead of a single north and a single south pole. Let us start with the solenoid. If we cut a coil in half, we would not be surprised that each half coil again would act like a bar magnet if a current flows through it (after some mild rewiring).

But a bar magnet is not made up of wound up wire. What corresponds in the case of the bar magnet to the wire carrying the current? It is the electrons whizzing around the nucleus in the atoms that provide the moving charges that generate the magnetic field. We are being rather simplistic here and the situation is a bit more complicated, it is not the electrons in the individual atoms, but groups of atoms that stick together. It is the current on the surface of these cluster of atoms, in a thin surface layer, that generates the magnetic field (the individual currents of the atoms inside the cluster cancel out). The bar magnet is made up of lots of these clusters that act like microscopic magnets.

If these tiny magnets align in an orderly fashion, then the individual fields generate a collective magnetic field on macroscopic scales, sufficient for a compass needle or a fridge magnet.

We discussed above that an electric charge gives rise to an electric field, and moving charges—a current—generate a magnetic field. Could "in return" a magnetic field generate a current? This is indeed the case, a changing magnetic field generates an electric current; the magnetic field has to change, similarly to the charges that have to move.

But the current consists of moving charges and each moving charge still has its electric field. Therefore the changing magnetic field leads to changing electric field, and a changing electric field leads to changing magnetic field, etc. These changing electric and magnetic fields will travel away from their source, which could be, for example, a wire having an alternating current flowing through it. This is another important effect of moving charges, to give rise to travelling electric and magnetic, or electromagnetic fields in short. We encountered this phenomenon before in Sect. 3.2.1: these travelling fields are nothing but electromagnetic waves. Figure 3.1 gives a nice impression of how the changing magnetic fields give rise to changing magnetic fields.

Observations indicate that on small scales and on large scales the universe is charge neutral, at least wherever we look these days, there are same number of positive and negative electric charges, and as far as we know, charge is conserved. For example, after walking across the carpet, generating a charge build up or imbalance, by touching the door we allow charges to flow from us back to the floor. There is charge neutrality afterwards, luckily for us as we experience the current from the charges, the small electric shock and possibly the spark, as rather unpleasant. But also on larger scales, we do not observe large scale charge separation and the resulting large currents, as these currents would have observational effects in the form of electromagnetic waves. The electromagnetic radiation we do observe has as far as we know different origins than some large scale charge imbalance, be it the Cosmic Microwave Background on the very largest scales, or nuclear processes on smaller scales in stars, and therefore galaxies.

Hence for cosmology on very large distance scales, only gravity is relevant, as although the electromagnetic force has infinite range just as gravity, the electromagnetic force has two charges that neutralise themselves, whereas gravity only knows one charge, mass.

Before we study gravity in detail, we have to discuss what we actually mean by "force" in the next section. This turns out to be not as simple a question as it looks at first glance, and also necessitates that we discuss the concept of "field" in physics in a little more detail.

6.3   What Are Forces and Fields?

Let us now have a closer look at the definitions of "force" and "field" that we already made casual use of in the sections above. Both concepts are probably already familiar to the reader in their classical interpretation, although the modern viewpoint might be surprising and take some time to accept.

In physics, a force causes an object to move, that is change its speed, or direction of movement, or change its shape, or a combination of these, unless opposed by an equal and opposite force. From this, rather dry, definition of force it follows that a force has both a magnitude or size, and a direction associated with it.

As an example, consider a book lying on a table. To slide it along the table in a direction will take a force in that direction to oppose the frictional force resisting movement, see Fig. 6.6. On the left of the figure the "experimental" set up, on the right the forces acting on the book. Let's assume the book is heavy and we begin by applying only a small force. Pushing at the book gently at first, nothing will happen, as the friction force prevents the book from moving. When we increase the pushing force, we will eventually overcome the friction force and the book will begin to slide across the table. We see that when the two forces are balanced, equal in magnitude and opposite in direction, the system will be in equilibrium, and in our example nothing moves, there are no changes in its current state of motion. We should also mention the other two forces acting on the book in Fig. 6.6, gravity or what we experience as the weight of the book acting downwards and the table providing an upwards force. These two forces are always in equilibrium in our experiment, or we would see the book suddenly moving vertically.

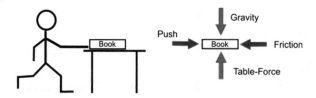

Fig. 6.6 The forces involved pushing a book lying on a table. On the left side, a sketch of our experimental setup. On the right, the forces acting on the book: gravity is pulling the book downwards (its "weight"), the table is supporting the book (force pointing upwards); we are pushing towards the right, the friction force is resisting our push, and therefore acting towards the left. If the friction force is larger than our push, the book just sits on the table, if we push harder and overcome the friction force, the book will move towards the right

We now turn to the question of how forces get transmitted and how we can describe this mechanism. This brings us back to the concept of "field" in physics. We already discussed some aspects of fields in the previous chapter, Sect. 5.5.2, and above in Sect. 6.2 for the case of the electromagnetic field, which we can now generalise.

The concept of a field is central to physics and to its applications. The idea is quite simple to describe but hard to imagine and visualise. A physical field is a quantity that has a value, or values, at each point in space and time. For the present we will assume that fields are smooth and well behaved.[2]

The simplest type of field describes quantities like the density or the temperature, which only require a single number to define them. The field then tells us the magnitude or size of the quantity, and where and when it has this size, that is at what point and at what time. The fields we are interested in here depend on time and the three spatial dimensions, or one time and three space coordinates. A field may be thought of as extending throughout the whole of space. In practice, the strength of every known field has been found to diminish with distance to the point of being undetectable.

For example, the temperature field describes the temperature at a particular time, and a particular position. This type of field is sufficient to describe, what physicists call, scalar quantities which can be described by a single number, like the aforementioned density and temperature, but also the pressure, or the amplitude of a scalar field.[3] The reader is probably familiar with temperature and pressure fields from the weather report on television, which often presents the temperature and pressure distribution across a region and how the distribution changes with time.

Staying with the example of the weather report, we see that we also need a directional field, or vector-field, to describe for example the wind speed. In this case, we need three numbers to describe the magnitude of the wind velocity in each of the three spatial dimensions. A vector field also depends on four coordinates to explain the when and where, but now instead of only one number, we need three numbers to specify the direction of the air flow, that is in what direction and how the strong the wind is blowing. We note that

---

[2] In principle, we should also worry about the region of space and the interval of time the field is supposed to be well behaved. It may of course be specified over all of space and for all time. Here we simply assume that the fields we are interested in are sufficiently well behaved over the regions of space and time we are interested in.

[3] The term "scalar field" is used precisely because these fields can be described by single number. The more mathematically inclined reader might recognise the scalar field as a function depending on time and the three spatial coordinates.

already in at first glance mundane applications, we encounter and make use of scalar and vector fields. Weather is surprisingly complicated!

Other examples of vector-fields are the electric and the magnetic fields we discussed above. The fields describe the magnitude and direction of the electric and magnetic fields in space and how they evolve in time. The fields also "communicate" the electromagnetic force between charged particles.

How can we visualise forces and fields? Representing a force graphically is not too difficult, we can use arrows to indicate the direction and the length of the arrow to indicate the magnitude of the force, as for example in Fig. 6.6. Graphically representing a field is more complicated. For fields that only represent a magnitude, that is scalar quantity like temperature or pressure, we can highlight regions that have the same value of the quantity. For example in weather maps points of equal pressure are connected, these lines are called isobars. This should help us visualising the pressure distribution in the region covered by the map.

We can also visualise fields of directional quantities. In this case it is usual to draw "field lines", these indicate the direction of a small "test particle" would take if placed in the field, and therefore also indicate the direction of the field. For example in Figs. 6.2 and 6.5 the field lines show the path a small electrical or magnetic charge would take. The density of the fields lines, how close they are together, indicates the strength of the field. Actually, we can make the magnetic field visible using iron filings. If we dust these onto a piece of paper lying on top of a bar magnet, the iron particles arrange themselves along the magnetic field lines.

Finally, returning to our example above in Fig. 6.6, where we might push the book with our finger. Also in this case the force get transmitted from the finger to the book with the help of the electromagnetic field. The molecules in our finger are held together by the electromagnetic field, and so are the molecules making up the book and the desk. When we push against the book, the electromagnetic fields of the flesh and skin molecules of the finger are interacting with the electromagnetic fields of the surface molecules of the book, thereby transmitting the force.

In the classical viewpoint on forces and fields described so far, forces are mediated by fields. However, our concept of what a force actually is has changed considerably in the last 120 years. Whereas it was introduced originally in Newtonian mechanics as a, at closer look, rather abstract concept, these days physicists view forces as interactions between particles, which are mediated by other particles, in the case of the strong, the weak, and the electromagnetic forces, or the force has been replaced entirely by geometry, as in the case of gravity.

This modern picture has been refined in recent years, and the fields communicating the forces have been replaced by the exchange of force carrying particles, see Table 6.1. In the table we see that the photon is the force carrier of the electromagnetic force, gluons mediate the strong, and the W and Z bosons the weak nuclear force. For example, a proton and an electron exchange photons and hence are attracted to each other; in the case of the strong force, the three quarks inside a neutron or a proton exchange gluons, hence they stick together.

Conceptually this picture of a force is quite intuitive. The more force carrying particles get exchanged between two particles, the stronger the force. Also the range of the forces follows neatly from this picture. The heavier the force mediating particles are, the shorter the range of the force. In the case of the weak and strong nuclear forces discussed above, the force carrying particles, gluons and W and Z bosons, are quite heavy (for elementary particles), hence the range of these forces is very small, making these forces relevant only on microscopic scales. The force mediator of the electromagnetic force, the photon, is massless (it has no rest-mass), and so the range of this force is infinite.

We discussed the particle-wave duality in Sect. 5.1.4. We introduced this duality discussing the electromagnetic field, and briefly discussed that it also can be extended to particles other than the photon. But it can also be applied to force fields and force carrying particles.

We should stress here that only our understanding and the concept of how forces are mediated has changed. The "character" of a particular force has not changed, gravity still attracts—more about this later in this chapter—and magnetic south poles attract magnetic north poles, only our understanding of how this comes about has changed. The strong and the weak force have been added in the twentieth century as an exchange of particles. Nevertheless, physicists still talk and think in terms of forces fields and force "attracting" or "repelling" particles and objects, for example. This can be confusing, but this is just a convenient short hand for the more complicated picture we just discussed.

However, gravity is different and doesn't actually exist in our modern point of view. Gravity has been replaced by the geometry of spacetime, as we shall discuss in the next sections.

6.4 Gravity, One Force to Beat Them All: On Large Scales

Gravity is the force describing how massive objects attract each other, and on large distance scales gravity is the dominating force in the universe. As discussed earlier in this chapter, the weak and strong nuclear forces have only a very short, finite range. The electromagnetic force does have an infinite range, but because there exist positive and negative electric charges, the effect of these charges cancels if there are the same number of each of them in the universe. On astronomical distances the universe is charge neutral, there are as many positive as negative charges overall and the electromagnetic force doesn't take effect. Gravity on the other hand has only one charge, mass, and therefore there can be no cancellation. All massive objects couple to, that is are affected by, gravity. Matters are "worse", as we shall discuss below, energy and mass are the same and even objects or particles that have no mass, or rest mass to be precise, but have energy like the photon are affected by gravity. Hence everything that has energy couples to gravity, and that is *really everything*. This is the reason why gravity plays such a crucial role in cosmology.

Before we discuss the modern view point on how gravity works, let us first take a look at Newtonian gravity, the "old picture" or concept of gravity.

6.4.1 Newtonian Gravity

In Newtonian physics gravity is described by a force law similar to the one of the electric force discussed in Sect. 6.2, although, as already mentioned, there is only one type of charge and the force is always attractive. Let us consider the simple case of two objects with masses m_1 and m_2 attracting each other, as shown in Fig. 6.7. The gravitational force experienced by the two objects is proportional to the product of the masses of the two objects, and inversely proportional to the square of distance between the two objects.

Fig. 6.7 Two objects with masses m_1 and m_2 are attracted to each other by gravity. The gravitational force is proportional to the product of the masses, and inversely proportional to the distance of the objects

Hence the magnitude of the gravitational force falls off very quickly, if the distance increases by a factor of 10, the force decreases by a factor of 1/100, but its range is nevertheless infinite. Therefore Newton's law of gravitation is another example of an inverse square law, as in the case of the electric force. The direction of the force is pointing along the line joining the centre of mass of each of the two objects. In Newtonian physics gravity is mediated by the gravitational field.

We can extend our simple example above to more realistic situations including many individual objects, such as the planets and their moons orbiting the Sun in the solar system, and also to objects more complicated in shape than the spheres in our example. In the case of the solar system we have to superimpose the contributions to the gravitational force of each object, the Sun, the planets and their moons. When calculating the gravitational force we can think of the mass of each sphere as concentrated at the centre of the sphere (the spheres behaving like point particles).

Similarly for more complicated shaped matter distributions, such as a blob, we can calculate the gravitational force by dividing up the blob into small volumes, or "volume elements", and then calculate the total gravitational force by adding up the individual contributions of each volume element. We treat each small volume element as an independent object and apply the gravitational force law to it.[4]

Gravity is always attractive but very weak, the constant of proportionality in Newton's force law, Newton's gravitational constant, is very small compared to, for example, the constant of proportionality in the electric force law (by 20 orders of magnitude). The effect of gravity is nevertheless dominant because the masses involved are huge, and as already highlighted, there is only one type of gravitational charge, mass, and therefore there is no cancellation or screening of the effect of the charge, as is the case for the electromagnetic force with its two charges.

At the end of the seventeenth century, Isaac Newton went beyond formalising the laws of mechanics and formulated what is now known as "Newton's law of gravitation". According to the, almost certainly apocryphal, story of the discovery of the law Newton was on an enforced holiday on the family farm owing to an outbreak of the plague. Sitting under a tree he noted an apple fall straight down to the ground and thought; the apple was attracted to the Earth and presumably the Earth was attracted to the apple. This idea was vital for

---

[4]This sounds rather complicated, but the mathematics behind it is straight forward. The technical term is to "integrate" or sum over all the individual contributions of the small volume elements.

arriving at his law of gravity that bodies attract each other with a gravitational force which is proportional to their masses and inversely proportional to the square of the distance between their centres. So the Earth attracts the apple with a force which depends on the mass of the apple, the mass of the Earth and the distance between their centres. The force is too small to have a noticeable effect on the Earth but once the connection between the apple and the tree is broken the apple accelerates as it falls towards the Earth. Newton realised that this observation is not limited to apples and the Earth but holds for all things everywhere.

Discovering this law and realising that it holds for *all* massive particles and objects was a remarkable feat. Gravitational fields exist across the whole of the universe and there is no way to avoid them or cancel them out because all bodies with mass are affected by gravity and there is no such thing as anti-gravity—all that happens if one adds more mass is to increase the gravitational attraction.

As mentioned above, on large distance scales gravity is the dominant force. On smaller scales, we are all familiar with its effects in our direct neighbourhoods: besides allowing us to pick fruit up from the ground that has fallen from a tree, gravity keeps people and objects on the Earth's surface, and it is Newton's law that governs the motion of the celestial bodies on all scales. Gravity keeps the moon orbiting around the Earth, and the Earth-moon system orbiting the Sun, together with the other planets of the solar system. But it also governs the motion of exo-planets, planets in orbit around stars other than the Sun, in their respective star systems. The inverse square law gives rise to the Keplerian motion of the planets around their stars discussed in Sect. 5.4.1. On larger scales gravity governs the motion of stars in galaxies, and holds clusters of galaxies together.

Newton's law works already very well in describing gravitational phenomena, and is sufficient in many situations when the distances and the masses involved aren't too extreme, such as in the solar system. However, as the accuracy and precision of the observations improved, astronomers noticed discrepancies between theoretical predictions of Newton's law of gravity, and the observational data as for example in the case of the perihelion drift of mercury's orbit, as discussed in Sect. 2.3.2.

This led physicists to seek a theory of gravity that could explain the observations, and resolve some fundamental problems of Newtonian gravity, such as how the gravitational field communicates the gravitational force between two objects. We will discuss the modern view on gravity in the next section.

6.4.2 Einstein's Theory of Gravity

At the beginning of the twentieth century it became clear to many scientists that the latest observations and experiments of the day were not in agreement with Newton's theory of gravity. Several physicists therefore began to develop a theory of gravity, or gravitation, that was consistent with observations, and consistent with modern concepts of physics, for example the non-existence of a special coordinate system or reference frame, which seems natural to us today but was novel idea at the end of the nineteenth century.

The scientist who first managed to pull all the new ideas and observations together and to come up with a consistent theoretical framework that would encompass the previous and very successful theory, Newtonian mechanics, but also allow for the new observational data to be explained was Albert Einstein. He published his work on, what would later be referred to as, the "special theory of relativity" or special relativity in 1905, and his work on gravitation and the "general theory of relativity" or General Relativity in 1915 and 1916.

We will introduce and discuss the key concepts and ideas of Einstein's special and general relativity relevant for cosmology below in the next sections. But let us first try and give a very brief overview.

Special relativity is build on the assumption that the laws of physics are the same in every frame of reference, or to be precise, in every inertial frame, that is a frame that is not accelerated. This implies that also the speed of light is the same in every coordinate system or inertial frame. General relativity extends and contains the special theory of relativity to allow also non-inertial frames, and formulates a new theory of gravitation.

6.4.2.1 A Cosmic Speed Limit

We begin this section, which deals with some key concepts of special relativity, with a short definition of what we actually mean by "reference frame" or "coordinate system" here. As mentioned previously in this chapter, we need three numbers to define a particular point or position in space, these are the three spatial dimensions or coordinates. In addition to where an object is or an event happens, we also need to know when the object is at this point or when the event takes place.[5] We therefore also need to specify the time, and we end up with four coordinates, time and the three spatial coordinates—up/down, left right, front/back, for example.

---

[5] An "event" in this context could be an observation being undertaken, or an experiment performed.

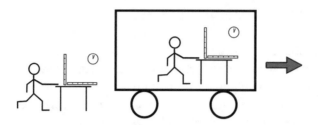

Fig. 6.8 To perform an experiment, we need a *coordinate system*, a set of three rulers to measure positions in space, and a clock. In this example, we measure the speed of light. On the left of the figure we measure the speed of light in a lab in a university building. On the right we perform the same experiment in a similar the lab on a train. Although the lab on the coach moves relative to the lab, we get the same result for the speed of light in both cases

We can think of the coordinate system as constructed from clocks and rulers to measure time and position. This situation is sketched in Fig. 6.8, where we draw only two of the spatial dimensions. We should use the term "reference frame" to denote the whole experimental setup, whereas "coordinate system" refers to the rulers and the clock, although usually we use the terms synonymously. We need the coordinate system to perform experiments, and to then evaluate our results. By contrasting the measured or observed experimental value with the predictions our theory made, we can test our theories as detailed in Chap. 2.

The experiment sketched in Fig. 6.8 measures the speed of light. Let us assume we perform the experiment in two laboratories, one is in a university building (on the left of the figure), the other similar lab is on a train which moves relative to the first lab with constant speed. In both cases we measure the speed of light to be exactly the same, 300,000 km, or 3×10^8 m/s. We should stress that the laboratories, or reference frames, move at constant speed, they are not accelerating. This type of reference frame is also called an *inertial frame*. Non-inertial frames, and accelerated coordinate systems, are the topic of General Relativity in the following sections.

Our experiment provides us with a first result, namely that the speed of light is the same in all inertial frames, whether they are at rest or whether they move at constant speed relative to each other. There is no special coordinate system that is preferred, all coordinate systems and inertial frames are equal and the speed of light is the same in all of them.

Let us assume next that we send a light ray from the lab on the train to the lab at rest, and measure the speed of the light ray in the lab at rest. We then reverse the experiment and send a light ray from the lab at rest to the moving lab and measure the speed of the light ray in the moving lab. For definiteness we take the train to move past the university lab that is at rest with a speed of 30 m/s (or 108 km/h), although the actual number is really not that important. In both cases we measure the speed of light to be exactly the same as before!

From everyday experience we would expect that if the train is approaching us and somebody sends a light ray towards us, the speed of the coach and the speed of light would have to be added together. This is similar to walking on a conveyor at the airport, where we have to add our walking speed to the speed of the conveyor, if we move in the same direction as the conveyor. But this is not the case! We measure the speed of light to have the same value, 300,000 km/s, in each of the cases described above, whether we are at rest or moving relative to the light source. The speed of light doesn't get larger even if we put the light source on a train, or a plane, or anything else fast moving. As we will see below, the speed of light is also the fastest speed that can be reached, and indeed represents a universal speed limit.

To make sense of our experimental results and to be mathematically consistent, we need to make several changes to Newtonian physics. We need to change the addition law for velocities, which then leads us to conclude that time and space are in fact no longer absolute and are connected, so that they have to be treated as a continuous four-dimensional spacetime, and also that mass and energy are related.

How do we add velocities? As mentioned above, in Newtonian mechanics, we can simply add velocities. For example, if we walk in the train coach in the same direction as the train, the speed with which we move relative to the ground will be the sum of the speed of the train and our walking speed. However, this is not strictly true, and only works for small velocities. Some fairly simple mathematical calculation allows us to derive an addition law for velocities that holds for all speeds, including very large velocities approaching the speed light. What we find is that in special relativity the law of addition is more complicated, but still contains the Newtonian "straight addition" law as a limiting case for small velocities. For large velocities we find that adding them up we get at most the speed of light, even if we replace the train in the above example with an extremely fast rocket. This is sometimes referred to as a "cosmic speed limit", namely that the speed of light is a speed that can not be exceeded.

To make sense of our experimental results, the cosmic speed limit has further consequences than complicating how we add velocities. We saw in Fig. 6.8 that we need to measure time and length scales to calculate the speed of light or the speed of any other object. It might therefore not come as a total surprise that the constancy of the speed of light also requires us to adjust the time and space dimensions, or time and the three spatial coordinates. Relative to an observer at rest, time passes more slowly for moving observers and they will observe clocks at rest going slower; this is referred to as time dilation. Time passes slower the faster the observer moves relative to the observer at rest, and the simple mathematical expression governing this relation also shows that the speed of light is again a limiting case, for an observer travelling at the speed of light time stands still. But also the length scales are affected. Moving objects contract along the direction of motion, the mathematical expression describing this length contraction is very similar to the expression for time dilation: the faster the object moves relative to the observer at rest, the more it contracts, the speed of light being again the limiting case in which the length of the object is zero.

The expressions and the effects of time dilation and length contraction are symmetric for the observer at rest and the moving observer, both measure the time being slowed down and length contracted in the case of the other observer. There is no special "coordinate system at rest", and therefore both can regard themselves at being at rest. It is impossible for either of them to perform an experiment to prove they are at rest and the other isn't.[6] We should stress that time dilation and length contraction are real, physical effects that affect the time and length dimensions or coordinates. They are not just simple rescalings as would be the case if we were to change the units of time or the units length.

In Newtonian mechanics time is absolute and independent of how an observer moves. We can also deal with observers and coordinate systems that move relative to each other, the transformation rule from the observer at rest to the moving coordinate observer and coordinate systems is simple, we only have to add the speed of the moving observer to spatial coordinates multiplied by time. The time coordinate itself is unaffected by the fact that the coordinates systems and observers move relative to each other. However, as outlined above, this leads to contradictions if we want to apply the transformation rules to coordinate systems and observers that move with large velocities, compared to the speed of light, or with the speed of light.

In special relativity, and also in general relativity as we will discuss in the next section, time is a coordinate similar to the three spatial coordinates and therefore also takes part in transformations. But time and space are no longer

---

[6]It might appear that in a situation as depicted in Fig. 6.8 we can decide which observer is at rest, but this is only due to the special experimental setup of laboratories being in s university and on a train. If both laboratories are on trains it becomes obvious that we can't decide which one moves and which one is at rest relative to the other.

independent from each other, they are intimately related and connected and form what is called, as already mentioned, the four dimensional spacetime. Time is no longer absolute, and will be different for observers moving relative to each other, depending on the state of motion of the observer. This is not as weird as it seems. We are used to "perspective", objects appearing larger or smaller depending on where they are, relative to the observer. Now, if we combine space and time and create spacetime, something we are led to by the speed of light being the same everywhere and coordinate invariance, we also have to allow for perspective effects in time. Just like spatial coordinates are subject to perspective changes, time also is subject to them, this is what we referred to as time dilation.

One way of looking at this is to allow for a clock and a set of rulers, that is a four dimensional coordinate system at every point in space. This might seem frivolous at first, but is even more important in general relativity as we shall see below.

The last implication of the cosmic speed limit that we need to consider here is that in special relativity also mass and energy are no longer independent. In order to discuss this mass-energy relation we need to first define more carefully than we have done so far what we mean by energy. In order to define energy, we need also to define work—in the sense used in physics. In Fig. 6.6 we have to expend work to push the book across the table. Work is simply the product of the force we have to use multiplied by the distance we push the book. For example, we need to do more work if we want to push the book further, and if we have to push harder. Energy is the possibility to do work. Energy is conserved and it can therefore not be produced, it can only be transformed from one form into another. Here we focus on two common forms, potential energy and kinetic energy, as shown in Fig. 6.9. An object has potential energy due to its position in a force field, as in the gravitational field in the example, or due to an electric field if it is electrically charged (see the discussion earlier in this chapter). An object possesses kinetic energy due to the movement of the

Fig. 6.9 Illustrating potential and kinetic energy. On the left figure a small marble is at rest at the rim of a large bowl, it has only potential energy. When the marble rolls down the bowl it picks up speed, in the figure in the centre, and converts its potential energy into kinetic energy. On the right of the figure, the marble has used up all its potential energy and only has kinetic energy. Not shown, the process in reverse when the marble climbs up the bowl, converting kinetic energy into potential energy

object, and the kinetic energy is equal to the rest mass of the object multiplied by the square of the velocity. The rest mass, as discussed above in Sect. 5.1.1, is the mass of the object that an observer at rest relative to it would measure. The reason for this definition will become clear below.

In the example of Fig. 6.9 a marble rolls down the inside of a large bowl. At the start, the marble only has potential energy. But as it rolls down inside the bowl it picks up speed, converting potential energy into kinetic energy. When it reaches the bottom of the bowl, it has used up and converted all its potential energy, and only has kinetic energy. This process would then take place in the reverse, and the marble climbing up the rim of the bowl would convert kinetic energy into potential energy.

We can now motivate why the cosmic speed limit also implies a relation between mass and energy. Let us return to the experiment in Fig. 6.6. Since it is a thought experiment, we can ask what will happen if we keep pushing for a very long time. If we keep accelerating the book we should be able to reach any speed we want, hence violating the cosmic speed limit. What is preventing this? One way to view this is that it's not mass but energy that we have to overcome, that is fighting its inertia, when accelerating an object. This link between mass and energy can be rigorously derived mathematically, but here it is sufficient to assume that the kinetic energy of an object contributes to its mass. In the actual mathematical expression we again find that the speed of light is a limiting case, that is the mass of the object grows without bounds as we reach the speed of light. We therefore can't accelerate any object with non-zero rest mass to the speed of light, let alone beyond it. This motivates that the cosmic speed limit also forces us to redefine our concept of "mass".

To conclude our discussion of mass and energy, we note that in the actual mathematical expression relating energy and mass, energy has two contributions, the rest mass energy and the kinetic energy. Mass contributes to energy, and energy is equivalent to mass—things can be massless, but not "energyless". If we can neglect the kinetic energy, we arrive at Einstein's famous relation, found nowadays on many t-shirts: *energy is proportional to mass, the constant of proportionality is the square of the speed of light.*

The cosmic speed limit also implies that only particles with zero rest mass can travel with the speed of light, for example the photon. It takes more and more energy to accelerate a particle with non-zero rest mass to high speeds, because the kinetic energy contributes to mass of the particle, making it effectively heavier the faster it gets.

It may seem odd to single out a speed as having special significance but it happens that the speed of light plays a fundamental role in physics. Particles can be accelerated to higher and higher speeds but not beyond the speed of light, it is the maximum speed that can be reached. The speed of light is a fundamental constant of nature, the same in all coordinate systems and set by the laws of electrodynamics.

We can generalise this and require that not only the laws of electrodynamics that determine the value of the speed of light, but that *all* laws of physics are independent of the coordinate systems and the motion of the coordinate systems relative to each other, if the relative velocity is constant. This is referred to as "covariance", the fact that there is no preferred coordinate system, and also requires that the speed of light is the same everywhere, even if coordinate systems move relative to each other. This is very different from physics before Einstein. As discussed above, in Newtonian physics we can just add velocities, even if they are large.

Finally we would like to stress that "relativity" in this context does not mean everything is relative! In fact the very opposite is true, it allows us to calculate precisely when things are happening, even for fast moving observers. It does mean however, that space and time are not absolute any more.

6.4.2.2   Geometry of Spacetime: The Rubber Sheet Analogy

We can now discuss the theory of gravity or gravitation which is at the heart of modern cosmology, Einstein's theory of general relativity. The essence of general relativity is that *the matter content of spacetime curves spacetime, and the curvature of spacetime tells matter how to move*, to quote J. A. Wheeler.[7] What does this mean?

We already introduced the concept of "spacetime" in the previous section. In special and in general relativity, we can no longer treat space and time as separate entities. Just as it would make no sense to separate out one of the three spatial dimensions and treat it separately from the other two, we now have to treat space and time together. The time dimension and the three space dimensions are still distinguishable and indeed behave very differently, but we have to combine all four into four dimensional spacetime, to construct a theory consistent with the experiments described above. The four-dimensional spacetime is dynamic, it changes with time and is also in space, it is elastic and can bend and deform. In Newtonian physics time and space, and therefore also spacetime are rigid and fixed, whereas in general relativity spacetime is elastic like a jelly.

---

[7]John Archibald Wheeler (1911–2008), American theoretical physicist, recognised for his work on general relativity.

Curvature is a familiar concept from standard geometry. We are used to think of curvature in a lower number of dimensions, often two, like surfaces embedded in our "normal" three-dimensional space. For example, the surface of a table is flat and is not curved, if it's a good table. On the other hand, the surface a sphere is curved, and has constant curvature. But we can easily think of more complicated surfaces, like the peaks and valleys in a landscape, or on the surface of an unmade bed. We can study these surfaces using mathematical tools, and using these tools it is straightforward to extend the concept of curvature to any higher number of dimensions. Although mathematically straightforward, curvature in more than three dimensions is difficult to visualise. Below we will however draw on our two-dimensional intuition and use two-dimensional analogies and examples to discuss the curvature of four-dimensional spacetime.

Finally, the matter content determining the curvature of spacetime includes all forms of energy. As discussed in the previous section, mass and energy are equivalent, and therefore any type of energy will contribute to the curvature of spacetime. This includes "normal" matter, like hydrogen gas, planets and stars, but also dark matter, electromagnetic radiation, and exotic fields.

The curvature of spacetime is usually very small, it is proportional to the energy content, but has as constant of proportionality the gravitational constant, which is itself small as discussed above, but also divided by the speed of light to the fourth power, making the constant of proportionality really very small! Therefore we need a lot of mass to end up with a sufficient amount of curvature to be noticeable. This is by no means obvious but built into the equations describing how curvature and the matter content are related.

The equations that arise from this picture or model, the governing equations of general relativity, are rather complicated, but can nevertheless be solved, either using pencil and paper—this usually requires some simplifying assumptions—or numerically, using high powered computers, as discussed in Sect. 2.5. Some of the quantities in the governing equations can be identified with similar ones in Newtonian gravity and are therefore still referred to as gravitational fields, and for this reason the governing equations are often simply referred to as "the field equations".

Since throughout this book we refrain from using mathematical equations, we have to resort to analogies to explain and motivate some of the key concepts

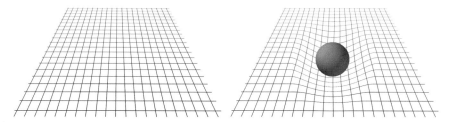

Fig. 6.10 On the left panel: empty four-dimensional spacetime represented by a two-dimensional elastic sheet. On the right panel we have a massive object bending spacetime, similar to putting a massive object on the elastic sheet, denting it

of general relativity in the following.[8] A picture may say more than a thousand words, but an equation says more than a thousand pictures.

Luckily there is a useful analogy: we can gain some intuition for curved spacetime by replacing the four-dimensional spacetime with a two-dimensional surface or sheet. We assume that this sheet is elastic and can be stretched, just like a rubber sheet or a piece of Lycra. We can place an object, for example a metal ball, onto the sheet and the ball will make a dent into the sheet. The heavier the ball, the deeper it will sink in, just like a heavier object will curve spacetime more.

We should however keep in mind, that this is only an analogy. First of all, we represent four-dimensional spacetime by a two-dimensional surface or sheet. Although the mathematics describing the curvature of four-dimensional spacetime and the curvature of a two-dimensional surface are similar, we should stress that general relativity deals with all of four-dimensional spacetime, and not just a two-dimensional surface. To be mathematically correct, we should represent the space part of the four-dimensional spacetime by a three-dimensional volume, this is however difficult or impossible to achieve visually, in particular if we then want to print the results in a book. We should also not think of objects somehow being "suspended" or "hanging" in the four-dimensional spacetime.

With these caveats in mind, the rubber sheet analogy works surprisingly well. In Fig. 6.10 we have on the left panel empty spacetime, represented by a two-dimensional rubber sheet. The sheet is not curved, as we have no masses included yet. What happens when we put an object, for example a massive

---

[8]Unfortunately it is much easier to solve the equations, than to use analogies and descriptions. This is usually not appreciated by non-physicists: physicists use equations because they are often much easier to handle than analogies and pictures!

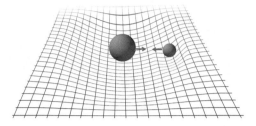

Fig. 6.11 Gravitational attraction in general relativity. The same setup as in Fig. 6.7 above, but now interpreted from a general relativistic point of view: two massive objects curve spacetime. The more massive object in the centre curves spacetime more than the smaller object. The smaller object to the right follows its shortest path in curved spacetime (similarly the larger object, but to a lesser extent). We still observe that the two objects move towards each other, and can still interpret this as an attractive force acting between them. However, they are just following their shortest paths in curved spacetime

sphere, on the sheet is shown on the right panel: the sphere makes a dent into the rubber sheet, it curves spacetime.

The curvature of the rubber sheet, the "dent", will be different given different mass distributions we put onto it. A marble that has a small mass will only sink in and dent the sheet a little bit, whereas a ball that has a larger mass will lead to deep dent or trough. In Fig. 6.11 we show the curved spacetime or rubber sheet due to two different masses. Also a small object that has a very large mass will lead to a very strongly curved sheet. Similarly, an object with a very large mass contained within a very small region will curve spacetime very strongly.[9]

We have now a better idea of what we mean by mass or energy curving spacetime. Let us now discuss how this the leads to, what appears as, objects experiencing a gravitational force. In general relativity, the concept of "force" is not replaced by an exchange of particles, as in the modern view point of the other forces, it is replaced by geometry. This is the geometry we discussed above, the curvature of spacetime. Instead of reacting to a force field, objects and particles follow the shortest possible paths in the curved spacetime. These paths of shortest distance are called geodesics and are the generalisation of the shortest distance in standard geometry, the straight line. However, here we have paths of shortest distance in four-dimensional, curved spacetime. All objects curve spacetime and then follow paths of shortest distance.

---

[9] A black hole is an extreme example where a large amount of mass is confined to a very small region of space. We will discuss black holes below in Sect. 6.4.3.

In light of this we can now view Fig. 6.11 with regard to the gravitational force in general relativity. The object with bigger mass in the centre curves the spacetime more strongly than the body with smaller mass.[10] In our setup, the small object initially is at rest relative to the large object in the centre, but it follows the shortest path in this curved geometry, moving towards the centre (when we let the smaller object move). We can still interpret this as the effect of an attractive gravitational "force", but the underlying, fundamental nature of this "force" is geometric, bodies moving on shortest paths through curved four-dimensional spacetime.

The field equations will also give rise to the well known inverse square law of Newtonian physics discussed in Sect. 6.4.1, and the apparent gravitational force is still proportional to the product of the masses, and inversely proportional to the distance of the objects. But this is only an approximation, valid when small gravitational forces are at play, as is the case for example on a laboratory on Earth.

Although we just explained that in general relativity there is no gravitational field in the classical sense (to mediate the gravitational force), we still find it a useful concept in our discussions. We should however bear in mind, that when we talk about the gravitational field and the strength of the gravitational field we mean the curvature of spacetime and its effect on objects and how strong this effect is. From a mathematical point of view, some of the quantities that describe the geometry of spacetime in the field equations of general relativity, can also be identified with similar quantities in Newtonian gravity, in particular the gravitational potential.

There is a simple relation between the gravitational potential and the energy density—the amount of energy per volume of space there is—in Newtonian gravity. In general relativity we can identify the quantities describing the geometry of spacetime with something similar to the Newtonian gravitational potential. We can therefore use the terminology and physical intuition from Newtonian physics here, and we will talk below also about the gravitational potential, keeping in mind again that it is a subtly different quantity in Newtonian gravity and general relativity. It is some times useful to think of a "gravitational landscape", which is due to the mass or energy distribution in the universe, which can also be described by potential wells and troughs.

---

[10]We can repeat this experiment using a rubber sheet, or using a Lycra cloth which is easier to purchase these days. Placing a larger mass body in the middle, and then a smaller mass body a short distance away inside of the dent made by the larger body, we will notice that the smaller body moves towards the larger one. Although the smaller mass body only follows the geometry of the rubber sheet and the larger mass body doesn't exert a noticeable force on the smaller one, we would interpret this as the larger body attracting the small one.

This would be similar to superpositions of Fig. 6.10, giving the appearance of an unmade bed. The more mass there is at a particular point, the deeper the potential well is.

In general relativity, just as in Newtonian gravity, there is only one charge, unlike in the case of for example electromagnetism. This charge is energy, and includes all types of energy, rest-mass energy (what we might naively think of as mass), kinetic and potential energy. Our experiment in Sect. 2.3.1 is therefore, with hindsight, not as crazy as it might have sounded at the time. Why there is only one gravitational charge is an interesting question, to which we still do not know the answer.

We discussed above that in special and in general relativity spacetime is a dynamic entity, it can change with time and in space. This is very different to Newtonian physics, where spacetime is fixed and rigid. We may think of it as just the stage where physics takes place. In general relativity spacetime is not just the stage where physics takes place, it is also a player. The quantities we can use to describe spacetime, the gravitational potentials, evolve in a similar fashion as the matter quantities, the evolving stage is part of the play.

Let us look at two more examples of objects experiencing the curvature of spacetime in general relativity, what we would usually refer to as gravity. In both examples we assume again that there are only these two objects in the universe (or at least the influence of all other objects is negligible). In Fig. 6.12 we see an object with small mass travelling through a spacetime which is curved by an object with a much larger mass. The small object is so much smaller in mass than the large object in the centre of the figure that we can neglect the curvature it imposes on spacetime. As above, the small object travels on a path of shortest distance in spacetime (a geodesic). Early on, before it is in the

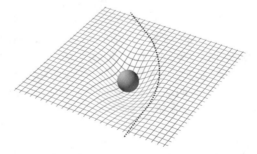

Fig. 6.12 The trajectory of a small object passing a massive object. The massive object curves spacetime, and the small object follows a geodesic, or shortest path in this geometry. For a light ray or photon we would have a similar trajectory, as light also follows the same shortest line or geodesic

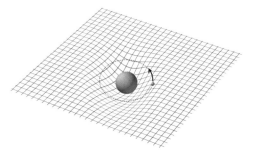

Fig. 6.13 Similar to Fig. 6.12, the trajectory of a small mass object passing a massive object, but here the small object is closer to the massive object in the centre. The massive object curves spacetime, and the small object follows a geodesic, or shortest path in this geometry. In this case the trajectory is different to Fig. 6.12 because closer to the massive object, spacetime is more strongly curved, resulting in a circular orbit of the small object around the large one

vicinity of the massive object, the small object travels on a straight line. But when it gets closer to the large object it enters the region of spacetime that is sufficiently curved to affect its trajectory and the path it follows will no longer be straight but is bent towards the massive object. To an observer it appears the small object is attracted by the large mass in the centre.

Since light and any other electromagnetic radiation also follow geodesics, or paths of shortest distance in spacetime, Fig. 6.12 also applies to light rays. This effect, the "bending" of light rays due to massive objects like stars and galaxies has been observationally confirmed.[11]

In Fig. 6.13 we have a similar setup as in Fig. 6.12, showing the trajectory of a small object with small mass passing a much more massive object, but here the small object is closer to the large mass in the centre. The massive object curves spacetime, and the small object follows a geodesic, or shortest path in this geometry. In this case the trajectory is different to Fig. 6.12 because closer to the massive object, spacetime is more strongly curved. Unlike in Fig. 6.11, we assume the small object has some initial velocity and moves as a result in a circular orbit around the object in the centre.

We already discussed in Sect. 6.4.2.1 that time for a moving observer passes more slowly than for an observer at rest. In general relativity there is an additional effect affecting the passing of time: gravitational fields also slow the flow of time. In the presence of masses, objects and observers move through

---

[11]The observation of light bending as predicted by general relativity in 1919 convinced most physicists that the theory was correct.

curved spacetime. This means, although they follow the shortest possible path in curved spacetime, the length of the path is longer than in a spacetime that is not curved, which is the case in the absence of masses and energy.

However, it is the length of the path in four-dimensional spacetime that determines the time that passes for a particular observer or object, this is also referred to as the "proper time" for the object. The curvature of *spacetime* means that gravity affects spatial dimensions *and* the time dimension. A clock in a gravitational field will tick slower, since it has to "travel" a longer distance in curved spacetime than if spacetime is not curved. This implies that time passes fastest for an observer at rest in an inertial frame.[12] In this case the length of the path only gets a contribution from the time part of the four-dimensional spacetime (there is no movement in space, only in time). Time also passes fastest, if the curvature of spacetime is as small as possible. This is in agreement with our lower dimensional intuition in two dimensions, the path between two points is shortest if we can go straight from one to the other, instead of following a curved path. Since massive objects curve spacetime, time will therefore also flow faster the less curvature there is, that is the smaller the gravitational fields are, or if they are completely absent.

Gravitational time dilation, the fact that time passes more slowly the stronger the gravitational field is, and hence the closer to a massive body, has been experimentally verified. A clock on a high tower will tick faster than a clock on the foot of the tower, because the gravitational field is weaker at the top of the tower (the strength of the field is inversely proportional to the distance from the mass). This effect is important, for example, for the Global Positioning System and other satellite navigation systems. Extremely precise clocks on these satellites allow users on Earth to calculate their position by measuring the time it takes a signal from a satellite to travel to the observer. Using three satellites then determines the position of the user in three-dimensional space. However, to get the required accuracy in the time measurements, the fact that the clocks are further away from Earth and therefore tick faster, but are also moving relative to an observer on Earth, and therefore run slower, have to be taken into account carefully. In the twenty first century, relativity has become a part of every day life!

The fact that time flows more slowly in the presence of massive objects, and hence gravitational fields, is often referred to as gravitational time dilation. A direct implication of this is gravitational redshift. As time flows slower and

---

[12] An inertial frame is a reference frame or coordinate system is one that isn't accelerating. In the presence of gravitational fields this can be achieved if the observer is freely falling.

the period of a wave is longer, the wavelength of an electromagnetic wave gets stretched, because we still require the speed of light to remain the same, as the speed of light is the distance travelled divided by the time passed. Hence we require a stretching of the wavelength to compensate for the slower clock and longer period, to keep the speed of light constant.

Finally, general relativity also gives the correct orbit of Mercury around the Sun, solving the perihelion problem we briefly discussed in Sect. 2.3.2. We should stress that Newtonian mechanics is not obsolete, it works extremely well in many settings, for example small gravitational fields and velocities much smaller than the speed of light, and only reveals itself to describe the universe less well than special and general relativity in extreme circumstances, for example high speeds or large masses, or when extreme precision is required, as in the perihelion problem. However, there are some problems it can't even attempt to tackle, such as gravitational waves, because Newtonian mechanics doesn't allow for a dynamic spacetime and therefore also can't deal with small perturbations or ripples to spacetime as we will discuss in the next section, Sect. 6.4.2.3. However, Newtonian physics works very well in many every day settings, since it is the low speed and low energy limit of special and general relativity.

Although we just discussed that in the modern view point gravity as a force doesn't exist, we nevertheless still talk about gravity as if it does. We use the term as a shorthand for the geometric view point, and objects still behave as if they are attracted to each other, although they simply follow shortest paths, or geodesics, in curved spacetime.

6.4.2.3 Gravitational Waves

Another surprising consequence of Einstein's theory of general relativity are gravitational waves. It took astronomers roughly hundred years until gravitational waves were directly observed for the first time on 14 September 2015 by the LIGO gravitational wave observatory. We discussed LIGO already in Sect. 3.3.4.2, but let us now discuss what gravitational waves are.

Gravitational waves are ripples in spacetime, as introduced in Sect. 3.2.2.2 when we discussed non-standard messengers in astronomy. We can get an idea of what this means if we replace the rubber-sheet in our lower dimensional analogy introduced above by the smooth surface of a pond. If we throw a small pebble into the pond, waves on the surface of the water will travel outwards from the point of impact. Similarly, gravitational waves are distortions of spacetime that travel outwards from their source with the speed of light, stretching and compressing spacetime periodically as they pass. As explained in Sects. 3.2.2.2 and 3.3.4.2, these periodic deformations of spacetime can be measured, or observed, using gravitational wave observatories. We can use a mirror as a "test-mass" by carefully suspending it in such a way that can it can respond to the periodic ripples of spacetime, like a cork bobbing up and down on the surface of a pond if a small ripple or wave passes it. As discussed in Sect. 3.3.4.2, we can then use interferometry to make the extremely tiny movements of the mirror visible.

However, this is just an analogy, gravitational waves are not surface waves, they are ripples of spacetime itself, periodic changes in the curvature of spacetime. If we think of spacetime as an elastic entity, like a jelly, gravitational waves are oscillations travelling through the jelly. Like water waves on the surface of a pond or electro-magnetic waves, gravitational waves oscillate perpendicular to their direction of motion, this is also referred to as being "transverse". This is by no means obvious, as for example sound waves are pressure oscillations in a medium, and the medium gets compressed along the direction of travel. Gravitational waves travel at the speed of light, that is $300,000$ km/s. Again, this is not obvious but follows from the theoretical framework and the equations of general relativity. Gravitational waves do not require a medium and can travel through empty space, like electromagnetic radiation. They also carry energy, like electro-magnetic and surface waves. A source emitting gravitational waves therefore looses energy.

Gravitational waves are generated when accelerating masses warp spacetime, leading to a sudden or a periodic change in the curvature of spacetime. The bigger the masses involved and the larger the acceleration, the stronger the gravitational waves emitted by the object or system. By stronger we mean here that the gravitational waves have larger amplitude, distorting spacetime more strongly, and carry away more energy.

If we accelerate a large mass suddenly we get a sudden change in the curvature of spacetime. A large mass corresponds to a strong curvature of spacetime, and if we somehow could accelerate this mass abruptly, we would get the closest analogy to the pebble being thrown into the pond, a sudden change in curvature travelling outwards from the point it was generated. It is however difficult to find a mechanism to give a large mass, for example a star, a sudden push and accelerate it. A more realistic example of how we can generate gravitational waves are two massive bodies orbiting each other, as we will detail below, or a single rotating massive body that is not spherically symmetric, wobbling about.

Let us illustrate the generation of gravitational waves by studying the case of two massive bodies, for example two black holes, in Fig. 6.14. The black

Inspiral

Merger

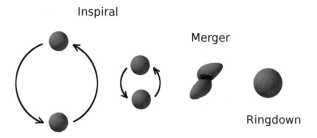

Ringdown

Fig. 6.14 Gravitational wave generation by two black holes orbiting each other. The orbit of the black holes shrinks as they loose energy by "churning up" spacetime and emitting gravitational waves. Towards the end of the process, the orbit shrinks rapidly and the black holes spiral inwards. Eventually the black holes merge and form a single, larger black hole. During ringdown the single black hole becomes more and more spherical, and settles down

holes orbit each other for thousands or millions of years, but their orbit shrinks because they loose energy by churning up spacetime and emitting gravitational waves. Towards the end of the process, the orbit shrinks rapidly and the black holes spiral inwards. As the distance between the two black holes decreases, their orbital speed increases. This effect is similar to a spinning ice-skater pulling in his or her arms, thereby increasing the rotation speed (this is due to the conservation of angular momentum). Eventually the black holes merge and form a single, larger black hole. Immediately after the merger this larger black hole is very non-spherical and rather blob like. However, during the final stage of the merger process, called "ringdown" the single black hole becomes more and more spherical, like a ball of jelly that has been squeezed but then returns to its spherical shape, while still emitting gravitational waves, and settles down. The last couple of orbits, the merger, and then the ringdown only take tenths of a second. Once the new black hole has stopped wobbling no further gravitational waves are produced.

From this discussion, we can already roughly predict how the gravitational wave signal due to this merger event will look like. During the final couple of orbits we expect the frequency of the gravitational waves, that is how often a wave peak reaches us per second, to increase. This due to the orbital speed increasing as the black holes get closer. Also we expect the amplitude of the wave, how much the wave distorts spacetime, leading to a displacement of the mirror, to increase until the merger. This is because the closer the two massive objects are, the stronger they distort spacetime. After the merger, during the ringdown, the amplitude decreases quickly, and the frequency of the waves remains roughly the same.

In principle any accelerated masses that curve spacetime sufficiently generate gravitational waves, but only if the masses involved are very large will this lead to gravitational waves that distort spacetime strongly enough for detectors on Earth to measure these distortions. Possible sources are therefore very massive accelerating objects, such as a pair of black holes or two neutron stars that orbit each other. Neutron stars and black holes are very massive objects, and in a tight orbit the acceleration the objects experience is immense. Another possible source of gravitational waves is the early universe. There we have no well defined objects like stars or black holes orbiting each other, but we have large density fluctuations moving around. Essentially we have blobs of denser material wobbling about which also will lead to gravitational wave generation.

Gravitational waves were one of the predictions of Einstein's theory of general relativity, published in a scientific journal in 1916. We discussed the LIGO gravitational wave telescope in Sect. 3.3.4.2 and highlighted how difficult it is to observe gravitational waves. It is therefore not surprising that it took nearly a century for them to be detected. Gravitational waves were observed directly for the first time in September 2015. We described the possible source of this gravitational wave event above and in Fig. 6.14. Let us take a closer look at what the LIGO detectors observed and measured.

Both LIGO observatories saw the mirrors in the tubes move in the distinctive way shown in Fig. 6.15, as a result of gravitational waves passing the detector sites. The strain pattern is consistent with the pattern derived from using computer simulations of the final orbits of two black holes, their eventual merger and ringdown.

The strain measurements of both observatories, showing the amount by which the mirrors were displaced by the passing gravitational waves, were not only in agreement with the predicted oscillation pattern (the predicted strain pattern is also shown in Fig. 6.15), but were also consistent with each other. The gravitational waves reached the Livingston detector 7 ms before they were observed in Hanford. When we shift the signal received by the Livingston detector by 7 ms and superimpose the two signals, we see from the bottom row of Fig. 6.15 that the two wiggly lines, the signal, are very similar. The distance between the Livingston and the Hanford sites is 7 lightseconds or 2100 km, hence the time lag between the two measurements is consistent with the gravitational waves travelling at the speed of light. The signals themselves show the increase in frequency and amplitude until the black holes merge and then a short period of decreasing amplitude, the ringdown, as we would expect after our discussion above.

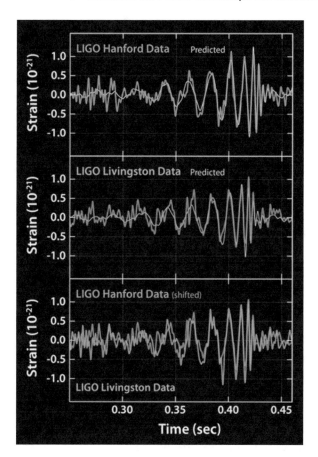

Fig. 6.15 The first gravitational wave detection or observation by LIGO on 14 September 2015. The plot at the top shows the data received at LIGO detector in Livingston, the plot in the middle the data from Hanford, and the plot at the bottom the two signals superimposed and shifted in time to allow for the distance between the two observatories. Time elapsing from left to right, the amount the mirrors get displaced by the gravitational wave relative to the length of detector arm, the strain, is measured on the y-axis (up or down). The peak strain measurements are at the time the black holes merge, at around 0.41 s. *Image Credit: Caltech/MIT/LIGO Lab*

Detailed computer simulations show that the source of this first gravitational wave observation were two inspiralling black holes with masses of 36 and 29 times the mass of the Sun. The single black hole that formed after the merger had a mass of 62 solar masses. The mass difference, three solar masses, was radiated away by the gravitational waves! For comparison, the Sun radiates during its lifetime less than 1% of its mass away in the form of electromagnetic radiation. During the final orbits and the merger, for tenth of a second, the energy per second was more than 20 orders of magnitude larger than the power output of the Sun. The distance to the event has been estimated to be roughly 440 million parsecs or about 1.4 billion lightyears from Earth.

This first gravitational wave observation, and the subsequent observations since 2015, are in beautiful agreement with Einstein's theory of general relativity, and further supports the theory. It also shows that gravitational wave astronomy has become a reality. Since the first detection in 2015 at the time of writing, 2019, ten more gravitational waves events have been observed. Besides the merger of black holes, also the merger of two neutron stars has been observed.

6.4.2.4   The Universe is Expanding

We discussed previously in Sect. 4.3 that Edwin Hubble found in the 1920s when he measured the spectra of distant galaxies that the spectra and the light that reaches us is redshifted. One possible explanation of this observation is that these galaxies move away from us, and when Hubble measured how fast they recede from us, he found that the further away the galaxies are, the faster they move away.

But if all galaxies move away from us, does this mean we here on Earth are somehow special? Are we in some sort of special place, the centre of the universe even? The short answer is no, we are neither special nor in the centre of anything. *Every* observer will see galaxies on large scales recede from her or him. This implies that the universe, indeed space itself, expands. But before we discuss the expansion of the universe further, we should take a step back and briefly discuss previous occasions in the history of astronomy, when we— humankind and Earth—were moved from the centre of things, or at least a special place, towards the sidelines.

The view of the universe has changed in the last couple of millenia quite considerably. By the third century BC Greek astronomers had established that Earth was a sphere, and Eratosthenes[13] calculated Earth's diameter surprisingly accurately, given the simple experimental and theoretical tools he had at his disposal at the time.[14] The model of the universe these Greek astronomers created had Earth at the centre, with the Moon, the Sun and the planets

---

[13] Eratosthenes of Cerene (ca. 280–195 BC), Greek mathematician and astronomer.

[14] Eratosthenes calculated the circumference of Earth by measuring the angle under which the Sun appeared at noon at Alexandria, Egypt, and at the same time in a town about eight hundred kilometres to the south, Syene. From the difference in angle, and the known distance between the two places, he was able to calculate the Earth's circumference.

moving around it. This is also known as the geocentric model and was completed in the second century by Claudius Ptolemy.[15] The Moon, Sun and planets were originally thought of moving around Earth on circles. To explain the movement of the planets which was the available data at the time, the model required however the introduction of epicycles or "circles on circles", the same motion a point on the side of a wheel describes when the wheel roles on the ground.

The geocentric model of the universe dominated cosmology until the beginning of modern astronomy in the sixteenth century with the introduction of the "heliocentric model" by Nicolaus Copernicus.[16] In the Copernican model of the universe Earth is no longer at the centre of things, it is a planet like any other, instead the Sun is at the centre.

Although Copernicus wasn't the first scientist to devise a heliocentric model, some Greek philosophers had similar ideas which however didn't become accepted, his move of Earth and its inhabitants "from the centre of things" to just an average place in the universe was nevertheless ground-breaking. These days Copernicus' insight has been generalised into the Copernican principle: we are not special, we are not living at a special place, everything is maximally boring or dull.

A problem of Copernicus' model was that it still required epicycles to model the motion of the planets, and was therefore quite complicated, making it difficult to use and its predictions not much better than the previous models. Using the latest observational data at the time that is the positions of the planets on their orbit around the Sun measured at a new level of precision with the help of the recently invented optical telescopes, Johannes Kepler[17] corrected the heliocentric model. He discovered that the planets aren't moving on circles, and circles on circles, around the Sun, but move instead on elliptical orbits. Here the relevant aspects of Kepler's laws are that the planets in the solar system move on elliptical orbits, and the speed of the planet increases until it reaches its closest point to the Sun, and it decreases until it reaches the farthest point. Using Kepler's laws astronomers were able to make much more accurate calculations and predictions of the positions of the planets, convincing the scientific community of the heliocentric model.

---

[15] Claudius Ptolemy (ca. 90–170), Roman astronomer who lived in Alexandria, Egypt; influential work on astronomy and geography.

[16] Nicolaus Copernicus (1473–1543), Polish astronomer, introduced the heliocentric model of the universe.

[17] Johannes Kepler (1571–1630), German astronomer and mathematician, discovered the laws of motion of celestial objects.

Kepler's laws are rooted in Newtonian physics, and hence are valid while Newtonian physics is a good approximation, that is for most applications when gravity is weak. However, a more accurate description requires the use of general relativity, when calculating the correct motion of Mercury as described in Sect. 2.3.2, or when calculating relativistic and strong gravity effects.

Although Earth had now moved away from the centre, the solar system was still at the centre of the universe. That also the solar system is not right in the middle of things, but part of and quite a distance away from the centre of a spiral galaxy, the Milky Way, was discovered during the further development of astronomy in the eighteenth and nineteenth centuries. However, astronomers thought that the Milky Way was the only galaxy, an island in an otherwise empty universe. We were still special. This last hold on a "special place" was given up at the beginning of the twentieth century, when observations showed that our galaxy is just one of billions of other galaxies in the universe (or possibly one of infinitely many!).

After this brief historical digression, let us now return to the expansion of the universe. When Hubble discovered in the 1920s that for observers on Earth and hence in the Milky Way all other galaxies are receding from them, it appeared at first that we might be in a special place after all, some sort of "centre of the universe" from which all galaxies move away. But there is no physical reason why this should have been the case. When we then invoke the Copernican principle, that is that all observers are equal, then all observers including the ones in distant galaxies, see the other galaxies move away from them.

How can all observers see galaxies moving away from them? The only possibility is that space itself expands. This is actually not as weird as it sounds, let us consider some bread dough with sunflower seeds, one of the standard analogies to illustrate the expansion of the universe[18]: the bread dough corresponds to space, the sunflower seeds to the galaxies, evenly distributed throughout the dough. If the dough rises and expands, an observer on one of the sunflower seeds would notice that all other sunflower seeds recede from him or her. But any observer on any of the seeds would notice the very same thing, all other seeds moving away. Again, observers might interpret this as them being in a special place, and all others moving away from them. However, this is obviously the wrong interpretation, as all seeds are equal and carried along by the rising and the expansion of the dough.

Luckily the expansion of space is easier to visualise than the curvature of spacetime, as the expansion affects three-dimensional space only, and not

---

[18] The recipe for bread with sunflower seeds is in Appendix A.5.

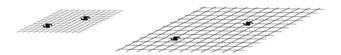

Fig. 6.16 Illustrating the expansion of the universe. Shown is a section of the universe including two distant galaxies, more than hundreds of million lightyears apart. On the left at some initial time, and on the right at a later point in time after the universe, or three-dimensional space, have expanded. The galaxies are farther apart, due to the expansion of space, the galaxies themselves didn't expand as they are held together by gravity (or spacetime curvature)

four-dimensional spacetime. Hence the number of dimensions we draw in the following figures correspond directly to the number of dimensions "in the real world". Let us start with a two-dimensional cross section of three-dimensional space in Fig. 6.16, to highlight the essential physics relevant for the expansion of space. We use two galaxies separated by several hundred millions of lightyears, the coordinate grid is "fixed in space". The expansion of the universe depends on time, on the left we see the setup at an initial time, on the right after some time has passed and the universe has expanded, in the example here by roughly a factor of two. That means that all distances in the universe, on large scales, have doubled. The galaxies themselves haven't moved, they have been carried along by the expansion.

We can also use a rubber sheet analogy similar to the one introduced earlier in this chapter to illustrate the expansion of space, and imagine that the coordinate grid in Fig. 6.16 is "painted" onto the elastic two or three-dimensional space, and the space gets stretched by the expansion. But as discussed, it is only space, the three "normal" spatial dimensions and not spacetime that expands.

Two questions usually arise at this point when introducing the concept of an expanding universe: what does the universe expand *into*, and does everything expand? Let us answer the second question first: not everything expands, space expands, but objects held together by forces, like elementary particles, atoms, people, planets, galaxies and clusters of galaxies, do not expand. The forces, either by exchanging force mediating particles or by curving spacetime, prevent the constituents of these objects from being carried along by the expansion of space. Nuclear particles, atoms and people are held together by the nuclear and the electromagnetic forces, whereas the solar system, galaxies and clusters of galaxies are bound together by gravity or the curvature of spacetime. Hence the expansion doesn't affect objects on scales up to tens of millions of lightyears, the typical size of galaxy clusters. But on very large scales, roughly above hundreds

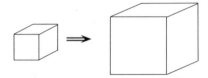

Fig. 6.17 Illustrating the expansion of the universe using three dimensions. On the left a large volume of the universe at some initial time. On the right, at a later point in time the universe or three-dimensional space, have expanded

of millions of lightyears, the expansion of the universe has observable effects, as measured by Hubble.

The first question can prove conceptually more difficult. The universe does not expand into anything, because the whole universe, all of it, expands. We haven't discussed the "shape" of the universe so far and whether it is finite or infinite in size, we will do so in Chap. 9. Suffice it to say here, that current observations point towards an either very large or an infinite universe. However, the size and shape of the universe isn't important when we discuss the expansion of the universe. What "all of space expands" means is shown in another simple figure. In Fig. 6.17 we show a large volume or three-dimensional section of the universe, with an edge length of the order of billions of lightyears, say. As in Fig. 6.16 we have on the left some initial time, and show on the right the same volume after some time has passed and the universe has expanded, in the example here by roughly a factor of two. If the universe is infinite, we could either imagine that the size of the box is also infinite, and all the universe is just the box, and the box expands. Or, we could leave the box as it is, large but finite, but let the universe be made up of infinitely many of these boxes, and all of them expand. Either way, there is no "into" into which to expand. The universe starts out infinitely large, all scales increase, the whole universe expands.

If the universe is finite, but in order to be in agreement with observations without a boundary, for example as in the case of the three-dimensional equivalent of the surface of a sphere, then all we have to do is take a finite number of the large boxes shown in Fig. 6.17. All of them will expand, and hence again there is no "other space" into which they could expand into. All of space expands, in this case we can visualise it by referring to the two-dimensional equivalent, the surface of a sphere. If the sphere expands, like a balloon filling with air, then also the surface of the sphere will expand. We should however not mistake the sphere, which is three-dimensional, with its surface, which is two-dimensional. In this analogy we would live on the equivalent of the surface of some higher dimensional sphere.

That space, and hence the universe, is expanding is not only logically and mathematically consistent, it is also in agreement with the governing equations of Einstein's general relativity when applied to the universe on large scales. Unlike a static universe, that needs somehow propping up to not be overcome by the equivalent of the gravitational instability which we discuss in the next section, an expanding universe is a stable solution. That a dynamic configuration, like an expanding universe, is stable whereas a static one isn't, is also familiar from every day life. Standing still on one leg is tricky, one can feel it is not stable. However, when we no longer try to stand still and restrict ourselves to be static but allow ourselves to move around, this changes quite dramatically, and hopping around on one leg feels much more stable than standing still.

Often regarded as an extension of the Copernican principle introduced above, namely that we are not at a special place or that all places are equal, is the cosmological principle, which states that all locations and all directions on very large scales are equal. We encountered the cosmological principle already in Sect. 2.4, and it can be expressed using more technical terms: on large scales the universe is homogeneous and isotropic in space, all locations and all directions are equal, in other words on large scales the universe is smooth and featureless. This is in agreement with observations, when we zoom out from any of the maps showing the Large Scale Structure of the universe we see no features beyond roughly several hundred million lightyears. On very large scales the dynamic of the universe and its matter content is described by the expansion of the universe, in agreement with the cosmological principle. On these scales, on hundreds of million of lightyears and more, where the universe is "simple" and smooth the governing equations are also surprisingly simple, and solving them is part of most astrophysics degree courses.[19] Nevertheless, the evolution of the universe on smaller scales is also important, but it can be messy and complicated, and the equations are much more difficult to study and to solve.

There is structure on smaller scales, on the scales of cluster of galaxies, galaxies, and planets. On these scales the universe is manifestly neither homogeneous nor isotropic. We can easily identify special locations, for example the position of a galaxy might be considered special compared to the empty space surrounding it, and also special directions, for example the direction along a filament in the cosmic web.[20] But on the scales relevant for cosmology,

---

[19]The variables in the governing equations in this case only depend on time, they become independent of position, which simplifies the calculation considerably.

[20]Galaxies are not randomly distributed, but on scales of hundreds of millions of lightyears form structures resembling a "web", see Fig. 4.9. We will discuss this further in Sect. 7.3.

we can "zoom out" from the scales of individual galaxies, clusters of galaxies, and individual filaments. On these cosmological scales, beyond hundreds of millions of lightyears, the universe *is* homogeneous and isotropic.

On large scales the dynamics of spacetime is described remarkably well by calculating how the distance between two far away points changes due to the expansion of the universe. Although on by cosmological standards small scales galaxies do move and change their position relative to each other, on large scales this motion and the change in position of the individual galaxies due to this motion is completely negligible compared to the change in position due to the expansion of space. The movement of the galaxies is often referred to as their "peculiar velocities", and in Fig. 6.16 would correspond to a small movement about the positions of the galaxies, too small to be seen in the figure. How the distance between two points changes *only* due to the expansion of the universe is governed by the "scale factor", which on very large scales is a function that only depends on time, which implies it has the same value at every point in the universe.

The scale factor also allows us to factor out the expansion of the universe on large scales, if we want: we can divide all length scales by the scale factor, which is also called using "comoving coordinates". This allows us to discuss the evolution of the universe using the positions of objects, such as galaxies, as they are carried along by the expansion of the universe. The coordinates in Fig. 6.16, determining the position of a galaxy irrespective of the expansion of the universe, are comoving coordinates. Comoving coordinates will simplify the discussion of the formation of structure on large scales in Sect. 7.3, and we already used them in Chap. 2 in Fig. 2.2.

Using the observations in Sect. 4.3 Hubble derived an expression that relates the distance of an object to its recession speed, which today is referred to as Hubble's law; the recession speed is proportional to the distance, the constant of proportionality is the current rate of expansion of the universe, also referred to as "Hubble's constant". The physical content of the law is that the recession speed of an object increases by roughly 70 km/s per million parsec.[21] That means two galaxies separated by 10 Mpc or 10 million parsecs would recede from each other with a speed of 700 km/s, if for the moment we ignore any other influences on them (such as gravity, or spacetime curvature). The further apart the galaxies are, the less they get influenced by the gravitational attraction of other nearby galaxies.

---

[21]Hubble's value was originally much larger, but the value of about 70 km/s per million parsec is in agreement with, and derived from, current observations.

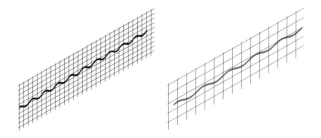

Fig. 6.18 Illustrating how the expansion of the universe leads to an increase in the wavelength of light and other electromagnetic radiation. We can "paint" a wave train onto an elastic sheet, and by stretching the sheet we see that the wavelength also gets stretched. On the left the original wave, on the right the stretched wave with now increased wavelength—redshifted

For example at the moment the Local Group of galaxies, including the Milky Way, and the Virgo cluster move towards each other with a speed of roughly 100 km/s. The two clusters of galaxies are roughly a distance 50 million lightyears apart. On the other hand, a galaxy cluster roughly a billion parsec or 3 billion lightyears away from the Local Group will move away from with a speed of 70,000 km/s, due to the expansion of space.

But the expansion of space not only affects the separation between distant points and objects like galaxies. The expansion also stretches the wave length of electromagnetic waves, such as light. This can be seen by using yet another rubber sheet analogy, as shown in Fig. 6.18. Let us assume we painted a single wave on an elastic sheet. If we then grab the sheet at its ends and pull, the sheet gets stretched and therefore also the wave gets stretched. The amount of stretching is proportional to the increase in wavelength. The expansion will affect electromagnetic radiation in general, including light which is simply radiation in the visible part of the spectrum (see Sect. 3.2.1 for a brief discussion of electromagnetic radiation). Because red light has a longer wavelength than blue light the stretching of the wave length due to the expansion of the universe is also referred to as "redshift". The expansion of space also affects gravitational waves in a similar fashion, that is the wavelength of a gravitational wave also gets stretched by the expansion of the universe.

The rubber sheet analogy works very well in this case, and detailed calculation shows that the redshift is directly proportional to the scale factor, which simply describes the evolution of the distance between points in the universe solely due to the expansion of the universe.

Therefore when we measure the redshift, or how much the wavelength has been stretched, of for example the light emitted by a distant object we measure different contributions to this effect: the stretching due to the expansion of space (as discussed above), the Doppler shift due to the motion of the object (as shown in Fig. 4.8), and a further contribution due to local spacetime curvature, see Sect. 6.4.2.2. On very large scales the expansion of space is the dominant effect. It can be difficult to distinguish between the different contributions, in particular on smaller scales.

Hence on large scales the redshift is directly related to the expansion of the universe. On smaller scales, "local" effects like the motion of galaxies due to the curvature of spacetime, or gravity, and the gravitational redshift directly related to the curvature of spacetime, make the interpretation of redshift difficult. Why is this important? We have established that redshift and the expansion of the universe, more precisely the evolution of the scale factor, are directly related. As already mentioned in Sect. 4.3, we can use the redshift of an object to measure its distance from us. Indeed, often cosmologists use the redshift measurements directly, instead of converting them into parsecs.[22] We will discuss this further in the final section of this chapter.

On large scales the expansion of the universe is the dominant dynamic effect. We should stress here again that by large scales we mean scales beyond tens of millions of parsecs.

Studying how the expansion of the universe changes over time, the rate of expansion, or whether the expansion of the universe decelerates or accelerates, is an important source of information about the constituents and the geometry or shape of the universe, as we will discuss below in Sect. 6.4.4.

6.4.3 Gravity Is Attractive: The Gravitational Instability

As discussed in the previous section, the dynamics of the universe, and therefore also how the matter content evolves, is on the very largest scales dominated by its expansion. But how does matter behave on smaller scales that are subject to gravitational attraction of other matter? This leads us also to one of the implications of gravity being only attractive.

Two objects attract each other gravitationally, something we discussed first from the Newtonian point of view, which assumes that there is a force as shown in Fig. 6.7 in Sect. 6.4.1, the force depending on the masses of the objects and their separation. But we now know that this view point is incomplete, and we therefore discussed gravity also from the more complete general relativistic

---

[22]For distance measurements only the contribution to the redshift due to the expansion of the universe is relevant.

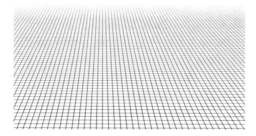

Fig. 6.19 An infinite grid, where for clarity we only show a two-dimensional cross section, but the grid could just as well be three-dimensional. At each intersection of the grid lines we put a mass. The distances between the masses are identical, also the masses are exactly the same. In this case the gravitational pull of all the masses balances exactly

viewpoint, in which there is no force, instead the masses curve spacetime and then move on the shortest path possible in this curved spacetime, as shown in Fig. 6.11 in Sect. 6.4.2. The important point here is however, that gravity is attractive.

In the absence of other forces, can anything stop the two bodies being attracted to each other, or counteract the attraction? The answer is no, because unlike in the cases of the other three forces discussed above in Sect. 6.3, there is only one type of charge in the case of gravity, and this "charge" is mass, or more precisely, energy. Hence there is no other charge, some "negative mass" that we could use to balance, counteract or repel gravity.

Let us now discuss some of the consequences of having just one kind of charge for gravity. It seems that if every thing is subject to gravity, and nothing can stop the gravitational collapse, we will end up in one single, big mass, after everything has been gravitationally attracted.

Maybe we can at least stop everything collapsing by very carefully arranging the matter in the universe. This is sketched in the thought experiment in Fig. 6.19, where we show the example of a two-dimensional cross section through an infinite three-dimensional grid, filling all of space. All the intersections of the grid are at exactly the same distance from one another, and at each intersection of the grid lines we put exactly identical point masses.[23] Since the masses and the distances are all exactly the same, and the grid and the distribution of masses is extending to infinity in all directions, all masses experience the same forces in all directions. The overall effect is that

---

[23]We can imagine the point masses to be small, massive spheres, of equal size and mass.

the gravitational pull on one mass due to all other masses cancels each other out (there are equally many masses to the left as there to the right of each mass). However, if one of the masses is only a tiny fraction heavier than the others, or is closer to one of the masses, the distribution of masses will become unstable. The slightly heavier mass will start attracting the masses surrounding it, or the mass slightly closer to another one will also break the balance. Once the balance is broken, the whole configuration becomes unstable: the slightly heavier mass or the masses closer to each other will start attracting more and more of the other masses in a run-away effect, accreting more and more mass. This run-away effect is also known as the "gravitational instability".

This means, even if we start out with a nearly perfectly smooth distribution of masses, if there is only a minute imperfection (as there always will be), the tiniest overdensity will start attracting more and more mass from its surroundings. Starting with a tiny overdensity, this overdensity will inevitably grow and "hoover up" all mass around it.

We introduced the term "overdensity" above which we haven't defined yet. By overdense we mean in this context that a region has a higher density than the average density, and "the overdensity" is the region denser than its surroundings. Similarly, an underdense region is a region of space that is less dense than the average.

Let us now look at slightly more realistic situations. To add realism to the thought experiment shown in Fig. 6.19, we have to take into account several effects that might influence the gravitational collapse, such as the "shape" of the overdensity, pressure forces and the expansion of the universe. A realistic description will need to include all of these effects, but we will attempt to discuss them separately as far as this is possible.

We begin with a discussion of how the shape of the overdensity, how the matter is distributed in space, affects the gravitational collapse. Consider an isolated overdensity of irregular shape which is made of pressureless matter. Let us assume that the blob is not completely irregular and amorphous but has the shape of an "ellipsoid", as shown on the left of Fig. 6.20, similar to

Fig. 6.20 Gravitational collapse of an irregular, non-spherical overdensity made of pressureless matter. We start on the left with an "ellipsoid" (all cross-sections of the blob are ellipses), the three major axes are all different in length. The blob collapses first along its shortest axis, and forms a flat, pan-cake like structure, as seen in the middle. The pancake can then collapse further, and we end up with the "filament", or a line-like structure, on the right

a potato. It is easier to discuss (and calculate) the gravitational collapse of an ellipsoid than of a completely irregular matter distribution. An ellipsoid is a generalisation of a sphere, and has three major axes of different length, all cross-sections of the blob are ellipses. A sphere is a special case of an ellipsoid, in which all major axes have the same length. We will discuss the collapse of spherical, more regular and symmetric overdensities below.

How the gravitational collapse of an ellipsoidal overdensity made of pressureless matter proceeds is shown in Fig. 6.20. We start on the left with a potato shaped ellipsoid. The blob collapses first along its shortest axis, and forms a flat, pan-cake like structure, as seen in the middle. The pancake can then collapse further, and we end up with a "filament", a long and thin or a line-like structure, on the right.

This process is particularly relevant on very large scales, of the order of hundreds of millions of lightyears (here we refer to scales today, since the universe expands and these scales therefore were much smaller earlier on, as we will discuss in Sect. 7.3, on the formation of large scale structure—we are using comoving coordinates, introduced earlier[24]). We already saw the gravitational instability at work on these scales in Fig. 2.2 in Chap. 2, Sect. 2.5. In this figure we start from a fairly smooth distribution of pressureless matter, with only small fluctuations in density, that grow through the gravitational instability, and turn into the cosmic web, with clearly visible filaments. On these very large scales the expansion of the universe begins to counteract the gravitational collapse, we end up with stable structures on large scales, as discussed in Chap. 7. On even larger scales the expansion keeps increasing the distance between regions of space and the matter contained in them, that in a non-expanding universe would have eventually collapsed.

Let us now move on to discuss the collapse of matter in the presence of pressure forces. We therefore focus on smaller scales than discussed above, and consider baryonic or "normal" matter and radiation.[25] We first consider the gravitational collapse of two smaller systems, where pressure forces play a role, but the size of the systems and the distances involved are still small enough to neglect the expansion of the universe. In Fig. 6.21 we show the collapse of a, by cosmological standards, very small volume. Its size is only a few hundred lightyears across, consisting of a fairly smooth gas cloud.[26] As discussed above in the context of the thought experiment in Fig. 6.19, there will be a region that

---

[24]By convention the scale factor is "one" today, hence today comoving and physical coordinates coincide.

[25]Radiation can behave like fluid if we have a very large number of photons with sufficiently large energy per volume, we will return to this topic in Chap. 7.

[26]We can think of the gas as hydrogen, but it doesn't really matter as long as it is made up of baryons.

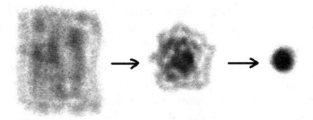

Fig. 6.21 Gravitational collapse. On the left we see a gas cloud, just a few lightyears across. Inevitably there will be a tiny overdensity in the cloud. The overdense region will attract more and more material, leading to a runaway effect. A blob forms, as shown in the middle of the figure. Eventually the collapse stops, when the gravitational force of the blob can't overcome the pressure force exerted by the gas. We are left with a nearly spherical much denser blob, as shown on the right

is slightly denser than its surroundings somewhere in the cloud, as there will always be small density fluctuations. This slightly overdense region will attract material, gas, from its surroundings become slightly more overdense, and so on. Again we have a runaway effect, the originally only minutely denser region will grow in mass. The region "hoovers" up more and more material from its surroundings. We end up with a much denser, fairly spherical blob. However, the collapse will stop, if the pressure force exerted by the gas compensates the gravitational force due to the gas in the blob. The blob will take on a spherical shape at the end of the process, as the material will move around until it reaches a configuration, or shape, of minimal potential energy. This shape is a sphere, as in this configuration every part of the blob experiences the same force.[27]

In larger volumes extending hundreds of lightyears up to millions of lightyears there will be not just one region that is slightly denser than its surroundings but several overdensities. Instead of collapsing into a single blob, the region separates into several smaller "subregions" or blobs that then can collapse independently, as described above. This is sketched in (Fig. 6.22).

As an example for gravitational collapse on scales of lightyears we can look back to the "Pillars of Creation" in the Eagle Nebula, in Chap. 4 Fig. 4.6. In this picture we see that some regions have already collapsed and formed stars. The pressure exerted by the radiation from these newly formed stars can

---

[27]To see that a sphere minimises its gravitational potential energy, just recall the example of the marble in a bowl. The marble looses potential energy rolling down, see Fig. 6.9. On the surface of a sphere all particles constituting the surface "rolled down" as far as possible, they are now all equally close to the centre of the sphere.

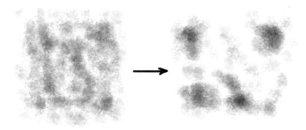

Fig. 6.22   A larger region of space than in Fig. 6.21. The region is large enough to contain several overdense regions. Instead of collapsing into a single object, the region separates into several smaller "subregions" or blobs that then can collapse

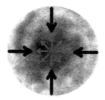

Fig. 6.23   The forces at work during gravitational collapse, using a spherical blob as an example. In this cross section the arrows indicate the forces and their direction: gravity, in black, pulling matter inwards towards the centre, in red we have the pressure force pushing outwards opposing the collapse

then be seen to drive away some of the material surrounding them, preventing further collapse.

Whereas it takes a blob of roughly a lightyear in diameter "just" a couple of hundred thousand years to collapse, on the largest scales of hundreds of millions of lightyears, the gravitational collapse takes billions of years. As we will discuss in the next chapter, the gravitational instability is still at work today and structures are still forming.

Let us return to the example of a smaller scale overdense region, as sketched in Fig. 6.21. What happens after the spherical, denser blob has formed? To answer this question, let us have a look at the relevant forces at work here, as shown in Fig. 6.23. The force working towards contracting the blob and compressing it is gravity due to the mass of the blob, acting inwards in the direction of the centre of the overdensity. Opposing the collapse is the pressure force exerted by the material of the blob. The further evolution of the overdensity depends therefore on the mass of the blob and its composition: if the gravitational force due to the mass of the blob is strong enough to overcome the pressure forces of the gas, then the collapse will continue. If the pressure force is able to counteract the gravitational pull, the gravitational collapse stops. The pressure force depends on the composition of the blob, what type of gas or mixture of gases, and their temperature. The gravitational collapse stops if the gravitational force is too small to overcome the pressure force, which can be the case if there isn't enough mass or energy in a region, either because the density isn't large enough and the region is too small, or if the pressure forces are too large.

The pressure force opposing the collapse cannot become arbitrary large, because as discussed in Sect. 5.3, pressure is the effect of particles interacting with themselves and these interactions are in turn governed by the forces discussed at the beginning of this chapter. We will discuss the implications of this in Sect. 7.1 in further detail,[28] and also what happens to the matter that undergoes collapse and gets compressed (spoiler alert: it heats up).

Furthermore, the pressure force itself contributes to the gravitational attraction. Another feature of the governing equations of general relativity is that they show that pressure also contributes to the gravitational attraction. This is not surprising if we bear in mind that in order to counteract the attractive force of gravity, which will compress the blob, we have to exert a pressure force. Keeping up this pressure force requires energy, but this energy also contributes to gravity (as discussed above in Sect. 6.4.2.1, energy and mass are the same). Hence if we require more pressure to prevent the collapse, eventually the pressure force will not counteract but contribute to the collapse. We can therefore never overcome the gravitational attraction leading to gravitational collapse by simply increasing the pressure, gravity always wins in the end.

Whether a region collapses, or not, depends on the size of the region, its constituents and composition, that is what stuff it is made of, and the properties of the stuff, for example the temperature and density of the gas. The size of the region and the density of the material in it determine the mass or energy of the region, therefore determining the spacetime curvature or gravitational force. The properties of the gas, its temperature, density and pressure, determine the pressure force and also the speed of sound of the material within the region, which controls how fast small disturbances in the gas travel, essentially telling us how fast the pressure can counteract the collapse. The result of this is that all regions above a threshold size will collapse, or in other words, collapse will take place on all scales above a certain

---

[28] Only the "non-gravitational" forces give rise to pressure, gravity itself is in most circumstances to weak to give rise to pressure-like effects.

scale. Studying the equations governing the gravitational collapse James Jeans[29] computed this threshold scale, called the "Jeans length". All regions or blobs larger than this scale will collapse, regions smaller than it will bounce back due to the pressure forces opposing the collapse and the region will oscillate in size. The Jeans length is proportional to the speed of sound and inversely proportional to the square root of the density of the region. This means the smaller the speed of sound and the larger the density, the smaller the Jeans length and hence also the size of the region that will collapse. The Jeans length is also affected by the expansion of the universe,[30] as the universe expands the Jeans length increases in direct proportionality.

How the collapse proceeds therefore also depends on at "what time" in the history of the universe we are, because this will not only determine the density and the dominating constituents of the universe, but also the rate of expansion of the universe and the evolution of the scale factor. We will discuss this below in the next section, Sect. 6.4.4. Because of these dependencies, during the different epochs of the universe the size of a region that will collapse due to the gravitational instability will change and can be very different from one epoch to the next. Hence also different types of objects will form at different times in the history of the universe and on different scales, if they form through the gravitational instability, as we will discuss in detail in Chap. 8.

To give a more concrete example of the gravitational instability at work on small scales let us discuss how an average star, like our Sun, forms and then continues to evolve, shaped by the interplay of the attractive gravitational force, and the opposing pressure force of the matter forming the star. We start out with an overdense region, slightly denser than the average density, of the order of light years in size, as sketched on the left of Fig. 6.21. The matter in the region consists mainly of hydrogen, but will also contain some helium and heavier elements (dark matter plays no role on these cosmologically small scales, see Sect. 7.1). As the cloud of gas collapses under its own gravitational force it gets denser, as shown on the middle panel of Fig. 6.21, and in the process heats up as the gas gets compressed. The pressure force in the interior is still not able to compensate for the gravitational force as the blob gets denser and denser. Eventually, as the gas further contracts, the pressure, density and

---

[29]James Hopwood Jeans (1877–1946), British physicist and astronomer, major contributions to cosmology and classical physics.

[30]The equations used to derive the Jeans length take the expansion of the universe into account, therefore the Jeans length is also proportional to the scale factor. As discussed above, only physical quantities held together by forces are not affected by the expansion of the universe. The Jeans length, since it is a theoretical construct, increases directly with the expansion.

temperature in the interior of the gas ball become so large that that the fusion of hydrogen to helium in the inner regions begins. This further contraction, until nuclear fusion starts, takes of the order of a million years in the case of a star like our Sun. The blob has become a star, the fusion reaction in its core then produces enough pressure to halt the collapse, the radiation produced in the process which also contributes to the pressure force, is the electromagnetic radiation we observe, once it reaches the surface of the star. The pressure force and the gravitational force, as shown in Fig. 6.23, are in balance with each other—remain in a stable equilibrium—, and the "hydrogen burning" continues for several billion years, in case of an average star.

Just to get a rough idea of the forces involved: the pressure in the core of the Sun is of the order 10^{10} or 10 billion bar, that is ten billion times normal atmospheric pressure on Earth, the temperature is of the order of 10^7 or 10 million kelvin. However this enormous pressure has to balance the gravitational force due to the mass of the Sun, roughly 2^{30} kg, to prevent further collapse.

At the end of the hydrogen burning period there is not enough hydrogen left in and around the core of the star to sustain the fusion reaction and produce enough pressure to balance the gravitational force of the star. As a consequence the core gets further compressed and the pressure and temperature increases. This then allows for other nuclear processes like the "burning" of helium into carbon to take place, for a relatively short period of the order of a few million years. During this period the star expands to become a "red giant" star, it then looses it outer layers and ends up as a very compact object called a "white dwarf". Although the white dwarf still shines, that is it emits electromagnetic radiation (light), it is no longer powered by nuclear processes. The radiation it emits is due to stored thermal energy, heat, from its previous "life". In the white dwarf the material got further compressed until only quantum effects prevent the core from further collapsing (the technical term is "electron degeneracy pressure", the electrons are forced much more tightly together with the nucleons than in the normal matter described in Sect. 5.1.2). How dense the material has become shows a comparison of the density of the core material of the Sun: the Sun has an average density of $1.4 \, \text{g/cm}^3$ (or $1 \, \text{cm}^3$ has a mass of $1.4 \, \text{g}$), the density of the Sun's core is roughly $100 \, \text{g/cm}^3$, whereas $1 \, \text{cm}^3$ of white dwarf material has a mass of $10 \, \text{mg}$ (or 10 metric tons). This has also implications for the size of the white dwarf. For comparison the Sun has a diameter of 1.39 million kilometres, whereas a white dwarf has a diameter of roughly 10,000 km.

The above sequence of events applies to an average star like our Sun, in the mass range of about 2–10 solar masses. The white dwarf can have a maximum mass of 1.44 solar masses, beyond that threshold mass even the degenerate electron pressure cannot prevent the star from collapsing further. This threshold mass is called the Chandrasekhar mass.[31] As already pointed out, the precursor star to the white dwarf is much heavier than the Chandrashekar mass, since during its lifetime the star looses mass by radiating energy, and also stellar material (in particular as a red giant).

But what about stars much more massive than the Sun? The further evolution again depends on the mass of the star. If the star was massive enough to end up as a stellar remnant beyond the Chandrasekhar limit but less than about 2 solar masses, then the material gets further compactified. In this case the pressure and temperature are so high that the electrons and the protons combine to form neutrons, and instead of a white dwarf we end up with a neutron star. We can think of the neutron star as a giant nucleus, having a diameter of the order of 10 km, but a mass of up about 2 solar masses. The material of a neutron is so dense that $1\,\mathrm{cm}^3$ has a mass of 10^{13} g or 10 million metric tons.

However, if the final stage of the collapse is even more massive than about 2 solar masses, then the gravitational attraction eventually overcomes even the more exotic mechanisms providing pressure, described above.[32] In this case we end up with a black hole.

Let us therefore briefly explain what a black hole is. We discussed in Sect. 6.4.2.2 how mass, or energy, curves spacetime. If we put more and more mass into a region of spacetime, the curvature it generates will increase. We show in Fig. 6.24 on the left a similar picture of spacetime as discussed previously, for example in Fig. 6.10, using again the rubber sheet analogy. However, here we omit to show the massive body located at the centre of the figure responsible for the curvature of spacetime. We assume that we have here a "normal" body such as a planet or a star, and hence the mass curves spacetime only gently. Particles or light rays passing the massive body at a distance get deviated. Also shown, a light ray or particle passing through the centre of the curved spacetime region: if it doesn't hit the object in the centre (which is small compared to the size of the curved region), it will simply pass through the curved region and leave on the other side.

---

[31]Subrahmanyan Chandrasekhar (1910–1995), Indian born astrophysicist, important contributions to theoretical astrophysics and stellar evolution.

[32]Here we allow for any mechanism that gives rise to *positive* pressure. We already discussed dark energy in Sect. 5.5.2, both the cosmological constant and scalar fields are popular candidates that can give rise to *negative* pressure. We will discuss this topic further in Chap. 9.

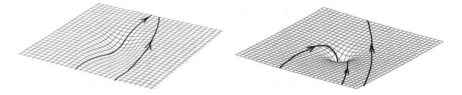

Fig. 6.24 On the left the familiar rubber sheet analogy, showing the curvature of spacetime due to a massive object (the object is not shown here). Two light rays, or particle paths, are shown. On the right: we added more mass but kept the volume the same, spacetime becomes extremely curved, a black hole forms. Nothing can escape, once it get too close to the central object and crossed the "horizon". Three light rays, or particle paths, shown, the two on the left ending in the black hole

If we now add more and more mass to the same region of spacetime, the gentle bump or shallow trough on the left panel will get deeper and deeper. In our rubber sheet analogy on the right of Fig. 6.24, we then have instead of a gentle bump a deep hole in the sheet, and eventually there is so much mass in the region that it curves spacetime so strongly that any object getting too close to the massive body will not be able to escape. What do we mean by too close in this case? The boundary of no escape around the black hole is called the "event horizon", and the distance from the horizon to the centre of the region is called the Schwarzschild radius.[33] Nothing can escape once it entered the horizon, not even light, hence the name black hole. The Schwarzschild radius is directly proportional to the mass of the black hole, that is the more mass we add the larger the radius. We should remind the reader that although we are only showing two spatial dimensions in Fig. 6.24, the horizon is a three-dimensional sphere. Particles or light rays passing the black hole at a distance will only get deviated from their original paths, as shown on the right of Fig. 6.24 (the deviation will be larger if the curvature is stronger).

We can also use the "escape velocity" to define the event horizon. The escape velocity is the speed we need to have reached to escape the gravitational field of a massive body, it depends on the mass of the body and the distance from the centre of the mass. For example, to escape Earth's gravitational pull, the escape velocity (on the surface) is 11 km/s, or roughly 40,000 km/h.[34] Because the speed of light is the maximum velocity anything can reach, as we discussed in Sect. 6.4.2.1 above, if we define a surface where the escape velocity reaches

---

[33] Karl Schwarzschild (1873–1916), German astronomer and physicist, early work on general relativity.

[34] The escape velocity is the speed at which the potential energy of an object is equal to its kinetic energy. See Fig. 6.9 and the discussion on energy relating to it.

the speed of light, nothing can escape from beyond this surface. The event horizon of the black hole consists of all the points where the escape velocity is the speed of light. This means nothing that crossed the horizon can come back.

A large mass in itself is not sufficient to form a black hole, in order to get a large spacetime curvature the mass has to be within a sufficiently small (comparatively) region. If we have a large mass spread out over a large region of spacetime, the resulting curvature will be too small to form a horizon. If we could compress Earth's mass into a sphere of less than 1 cm radius, we would end up with an Earth-mass black hole, whose Schwarzschild radius is roughly 1 cm. Similarly, compressing the Sun sufficiently, we would get a black hole with an event horizon of 6 km in diameter, the Schwarzschild radius in this case is roughly 3 km. The Milky Way, consisting of roughly 10^{11} or one hundred billion stars and a much larger amount of dark matter, has a total mass of about 10^{12}, or one thousand billion solar masses. If we were to compress all this mass into a small enough region we would get a black hole with an event horizon of about 2×10^{12} km or about 0.2 lightyears radius. If the mass is very big, as in the case of our galaxy, the resulting black hole will also be "astronomical" in size.

As discussed above, black holes are the final stages in the evolution of stars with masses above about 20 solar masses (after exploding, their remnants are more massive than 2 solar masses). However, they can also be formed if enough mass is concentrated in a small enough region of space, or, put in other words, if the curvature of spacetime is large enough to form a horizon. This may happen in the early universe, where fluctuations in the density of the matter present at very early times can be sufficiently large. We will revisit this in more detail in Chap. 9, where we discuss the "initial conditions" that is the beginning of the universe.

Before we discuss the evolution of the universe in the next section, we should point out that Einstein's theory of gravity is non-linear. In a linear theory, for example, if we double one quantity a dependent quantity will also double, it will increase in "linear proportion". In a non-linear theory the output might depend in a complicated way on the input, for example quadratically or exponentially. Performing calculations in a non-linear theory is much more difficult and complicated than doing sums in a linear theory. Luckily, in many settings relevant to cosmology we can treat gravity as being close to linear, which simplifies the calculations considerably. This is for example the case when the masses involved are spread out over a large region or volume, and therefore the mass- or energy density relevant to the problem are small, and the errors we will make by using linear theory are also small. The early stages of structure formation through the gravitational instability can be treated linearly, while the densities involved are still small. At later stages in the structure formation process, when the densities have grown, we have to use the full non-linear theory. Also, if the masses involved are large and the region is comparatively small, and therefore the spacetime curvature is also large, then gravity is very non-linear, as for example in the case of two merging black holes, as shown in Fig. 6.14. In this case calculations are extremely complicated and require the use of large super-computers.

The difference between the linear Newtonian gravity and general relativity, which is non-linear, becomes clearer when we focus on the gravitational fields and, unfortunately, the governing equations, as this difference manifests itself in the governing equations. To be precise, the general relativistic equations are non-linear in the gravitational fields. If the masses involved are small, or the density—the mass per volume—is small, then the non-linearities in general relativity do not become apparent. Hence Newtonian gravity works so well in this case, for example when applied to the orbits of the planets in the solar system.

If the masses involved are large, then also the spacetime curvature and hence the gravitational fields are large. In this case we can't simply superimpose gravitational fields as in the Newtonian case, this only works when gravity is weak (and linear). In the strong field case, gravitational fields don't simply add up linearly, the fields are nonlinear (doubling the field strength more than doubles the curvature or force).

The non-linear effects of gravity also come to the fore when we need very accurate results. In this case, even if the gravitational fields involved are fairly weak, we have to use general relativity instead of Newtonian gravity. We already encountered an example for this in Sect. 2.3.2, when we discussed the perihelion shift in the orbit of mercury.

Let us highlight the salient points of this section. Gravity is attractive, hence in a smooth distribution of matter, for example some gas, a tiny overdensity will attract more matter and therefore gain more mass. This leads to an even larger attractive force, or stronger spacetime curvature, and attract even more matter, leading to a runaway effect, attracting more and more mass. The gravitational instability leads through the growth, and therefore amplification, of tiny density fluctuations generated in the early universe on very small scales to the distribution of galaxies and clusters of galaxies on the largest scales. On smaller scale it is also responsible for the formation of galaxies through the collapse of giant clouds or blobs of dark matter and hydrogen, and on even smaller scales for the formation of individual stars and planets. In a sense, gravity is not only responsible for the formation of the solar system, it makes the Sun shine by providing the high pressure and temperature through to force hydrogen atoms to fuse into helium.

6.4.4 How Old Is the Universe, and How Big?

Before we move on to discuss how structure forms in the universe in the next chapter, let us conclude this chapter with a discussion of the age of the universe and its size, both closely related to the expansion of the universe. When we talk here about the size of the universe, we have in mind the size of the *visible* universe. We will discuss the overall size of the universe later in the next chapters.

We saw in Sect. 6.4.2.4 that observations led us to conclude that the universe is expanding, and that on large scales all points, or places, in the universe move away from each other at the same rate, as illustrated in Figs. 6.16 and 6.17, we introduced the scale factor which describes how on large scales distances change due to the expansion of the universe.

In a expanding, dynamic universe, governed by Einstein's theory of gravity we do not have the luxury of absolute time any more, as we would in a universe governed by Newtonian gravity. We discussed above that in general relativity time and space are no longer separate and we have to deal with them together as four-dimensional spacetime. Energy, or mass, curves spacetime, and the time measured by an observer depends on how strongly spacetime is curved.[35] This also complicates the calculation of the age of the universe. However, we can use the expansion of the universe and the scale factor, which describes on large scales how spacetime evolves, also to calculate its age. This is by no means obvious, but can be deduced from the governing equations.

Let us assume the universe kept expanding from its beginning. At early times the scale factor will have been very small, approximately zero, and we can use this value to define the "time at the beginning", and the scale factor today to define "today". But to make use of this, we have to also know how the scale factor evolves in between these two points in time, and the next step is therefore to calculate the evolution of the scale factor; this depends on the matter content of the universe, through the governing equations. We already saw that on small scales, by cosmological standards, gravity manifests itself through the gravitational instability, giving rise to the structure we observe

---

[35]The curvature of spacetime is equivalent to the presence of gravitational fields, and a clock in a gravitational field will tick slower than a clock very far away from the source of the field, where the field is weaker.

in the universe. On much larger scales, beyond hundreds of millions of light years, gravity controls the expansion of the universe. The speed, or rate, of expansion, how the scale factor changes with time, is also known as the "Hubble parameter", and its value today is referred to as Hubble's constant, as discussed in Sect. 6.4.2.4 when we introduced the expansion of the universe. On these very large scales, the total energy density directly controls the rate of expansion (the Hubble parameter is proportional to the square root of the total energy density). All types of matter contribute to the energy density, hence as long as the energy density is non-zero also the rate of expansion will be non-zero and the universe keeps expanding, although the rate of expansion will vary depending on the energy density. However, how the speed of the expansion changes with time, which can be expressed as the time rate of change of the Hubble parameter, and hence whether the expansion accelerates or decelerates does depend both on the amount *and* the type of matter present in the universe. All "non-weird" matter—radiation, normal or baryonic matter, and dark matter—will lead to the rate of the expansion slowing down, the expansion of the universe goes ahead ever more slowly. Only weird matter, for example dark energy, can give rise to a speed up of the rate of expansion, to which we usually refer to as an accelerated expansion.[36] Let us therefore have a look at the evolution of the different types of matter which we already encountered in previous chapters. In particular we can quite easily find the evolution in terms of the expansion of the universe, that is in terms of the scale factor.

The behaviour of different types of matter in an expanding universe is another direct consequence of Einstein's theory of gravity, and the governing equations derived from it. Although we would need to use the governing equations to calculate how these quantities evolve in a precise, quantitative way, we can use some of the ideas and physical concepts presented in this and the previous chapter to get an idea of how these quantities behave.

We discussed the main constituents of the universe that are relevant for this discussion, radiation, matter and dark energy, in Chap. 5. On large scales they respond very differently to the expansion of the universe and get diluted at different rates, and hence the composition of the universe, the contribution of these ingredients to the overall energy budget of the universe is different at different times. Here we can group normal matter and dark matter together,

---

[36]We should stress here again, that the relations between the expansion rate and the energy content of the universe, and the rate of change of the Hubble parameter and energy content, are by no means "obvious". They follow from the governing equations of Einstein's theory of gravity.

since on the large scale we are interested in they behave similarly, both can be treated as pressure-less fluids.[37]

The quantities we are interested in here are the energy densities of the constituents, that is the amount of energy in each constituent per volume. We begin our discussion with matter that can be treated as particles. This includes dark matter, and "normal matter" on sufficiently large scales, either in the form of diluted hydrogen and helium, or in the form of individual galaxies. In either case the particles—atoms or galaxies—are sufficiently separated from each other that they cannot interact with each other and therefore can be treated as a pressureless fluid, or in the terminology of cosmologists as "dust".[38]

With this description of matter in mind, let us consider a large cubic volume containing a fixed number of particles. As the universe expands, the cube will also expand as shown in Fig. 6.17. Because the number of particles is fixed and remains therefore the same, and the volume has increased, the number per volume will decrease. We can be more specific, since the volume of the cube we chose is the product of its three edges and each edge will increase with the expansion of the universe in direct proportion to the scale factor, like all length scales. Therefore the increase of the volume is proportional to the cube of the scale factor, and the matter number density, the number of particles per volume, gets diluted like the inverse volume or $1/(\text{scale-factor})^3$, the inverse cube of the scale factor. If we assign each particle an energy, for example due to its rest-mass, we find that the energy density of matter is inversely proportional to the volume, that is inversely proportional to the cube of the scale factor on large scales, or $1/(\text{scale-factor})^3$.

Next we take a large cubic volume filled with electromagnetic radiation, and consider how the energy density of the radiation responds to the expansion of the universe. The energy density of the radiation gets diluted faster than that of the pressureless matter, because we now not only have to take into account the volume increase, but also that the wave length of the radiation gets stretched by the expansion. Because the energy of a photon is proportional to the frequency of the electromagnetic radiation and inversely proportional to the wavelength, the combined effect of the volume increase—a dilution by a factor of $1/(\text{scale-factor})^3$—and the increase in wavelength—contributing a factor of $1/(\text{scale-factor})$—is that energy density of radiation decreases as

---

[37] As discussed in Chap. 5, pressure is a small scale phenomenon arising from the interaction of particles. The interaction scales for normal matter are far too small. For radiation, owing to the infinite range of the electromagnetic interaction, we do need to take into account its pressure.

[38] Specks of dust can be seen floating in the sunlight, are not interacting with each other, providing a neat analogy for galaxies "floating" in space.

$1/(\text{scale-factor})^4$, the inverse of the scale factor to the fourth power. Radiation gives rise to pressure and it is therefore not surprising that radiation responds differently to the expansion of the universe than a pressureless fluid.

Since the result above applies to any electromagnetic radiation on large scales, we can also use it to find how the temperature of the Cosmic Microwave Background changes as the universe expands. The energy density of the Cosmic Microwave Background radiation gets diluted as the inverse of the fourth power of the scale factor. Because the Cosmic Microwave Background radiation has the spectrum of a black body, its energy is completely determined by its temperature, and the energy density is proportional to the fourth power of the temperature. This is—again—not obvious but can be seen from the governing equations. However, equating these two relations we can see how the expansion affects the temperature of black body radiation: the temperature is inversely proportional to the scale factor, therefore the universe cools as it expands. If we run the expansion backwards, we find that the universe was hotter when it was younger. We will return and make use of this fact extensively in the next chapters.

The last ingredient we need is dark energy, which we have discussed in Sect. 5.5. Dark energy doesn't get diluted at all, and is also in this respect very different to the other contributions to the total energy budget. Let us use the simplest example of dark energy, the cosmological constant, as an example to motivate this unusual behaviour. The cosmological constant is *defined* to be constant throughout the universe. That means at every point in space we would measure the same, tiny value for the cosmological constant, or in terms of energy density, all of space is filled with the tiny energy density due to the cosmological constant; the energy density of the cosmological constant is spread through all of space and does not change. The fact that it doesn't get diluted stems from its definition, and it therefore has also some unusual physical effects, such as a negative pressure.

We can now plot how the energy densities of the three main contributors to the total energy budget of the universe evolve. In Fig. 6.25 we plot the energy densities of matter, which includes both dark matter and baryonic matter, radiation, and dark energy relative to the total energy density. We use here redshift as the time coordinate on the bottom and give values for the equivalent standard time at the top of the graph.[39] Since we give on the vertical axis the energy densities relative to the total energy density of the universe, the

---

[39] Redshift is here, and elsewhere in cosmology, a convenient time coordinate, as it is directly related to the expansion of the universe and directly measurable, as discussed previously.

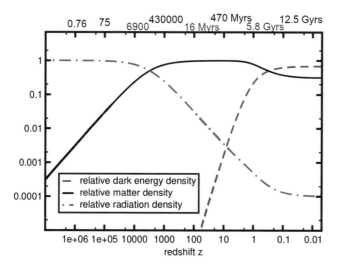

Fig. 6.25 The energy densities of matter (black), radiation (red), and dark energy (blue) relative to a reference density (the total energy density). We use the simplest model for dark energy, a cosmological constant. Time increases from left to right: the figure starts at the very left at redshift 10^7 (10 million) or roughly 66 h after the beginning, and ends at the very right at redshift 0 or 13.8 billion years after the beginning. At the top of the figure the time in years (see also Table 6.2)

values add up to unity, or 100%. We use values for the energy densities as discussed in Chap. 1 and shown in Fig. 1.2, and at the present time there is roughly 69% dark energy, 30% matter (25% dark matter, the rest "normal" baryonic matter), and a tiny fraction of radiation (0.01%). Time increases on the bottom axis from left to right, starting at the very left at redshift 10^7 or 10 million which corresponds roughly to 66 h after the beginning, and ends at the very right at redshift zero or 13.8 billion years after the beginning. The numbers at the top of the figure give the time in years (Myrs and Gyrs stand for millions and billions of years, respectively).

We see from Fig. 6.25 that at different times different components dominate the energy budget of the universe. At the beginning radiation is the dominant constituent, until at redshift 3400, or roughly 50,000 years after the beginning, matter takes over. This switch in dominant component, the time when the energy densities of matter and radiation are equal, is also known as "matter-radiation equality". Matter domination continues until about redshift 0.31, or when the universe was 9.3 billion years old. From this point in time, when the energy density of matter and dark energy are equal, dark energy begins to dominate the evolution of the universe, and we live now in this dark energy dominated epoch. We should stress, that all three constituents are present

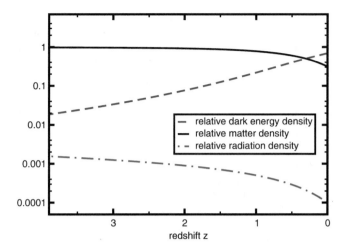

Fig. 6.26 The relative densities of the constituents of the universe. Same as Fig. 6.25 above, but zooming in on the last 10 billion years, using a non-logarithmic time axis at the bottom. Here we can clearly see the turn-over from matter to dark energy domination at redshift of about 0.3, when the universe was roughly 10 billion years old

throughout the three epochs sketched above, but only one is dominant, the other two are subdominant or negligible, as can be seen from Fig. 6.25. How redshift, time, and scale factor are related is also shown in Table 6.2.

The evolution of the mixture of matter, radiation, and dark energy is shown again in Fig. 6.26. Unlike Fig. 6.25, where we used a logarithmic time axis (showing multiples, or powers of 10) which allows us to "bunch" 13.8 billion years into the figure, we use a different scaling for the time axis, starting at redshift 4, when the universe was 1.5 billion years old, therefore showing only the matter and dark energy dominated epochs. We arrived at Figs. 6.25 and 6.26 by solving the governing equations, using the relations discussed above. We also needed the time evolution of the scale factor, as described next, which can then be related to the redshift.

We now have all the necessary ingredients to calculate how the scale factor, describing the expansion of the universe, evolves. We mentioned above that the expansion rate of the universe, the time rate of change or how fast the scale factor changes with time, depends on the total energy density of the universe. But the total energy density is simply the mixture of matter, radiation, and dark energy that we just discussed, and it is just a short calculation to find how the scale factor changes with time.

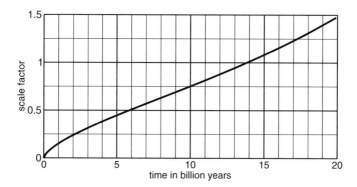

Fig. 6.27 The time evolution of the separation of two distant points in space, also known as the scale factor of the universe. On the horizontal axis time in billion years, on the vertical axis the scale factor. The scale factor is chosen to be unity today, 13.8 billion years after the beginning. The distances in the past relative to distances today were smaller, for example 6 billion years after the beginning distances were just half the size they are now

The result of this calculation, the evolution of the scale factor, is shown in Fig. 6.27. We see that the scale factor grows throughout the history of the universe. The behaviour of the scale factor during radiation domination, roughly the first 50,000 years, and matter domination, roughly the next 9 billion years, is quite similar. We therefore cannot see the change in growth from growing proportional to the square root of time during radiation domination to growing at a slightly faster rate, namely as the cubic root of the square of time, during matter domination in the figure. The change from matter domination to dark energy domination is more dramatic and can also be seen in the figures. Once the universe is dark energy dominated the scale factor begins to grow exponentially. There is nothing special about the point in time in the evolution of the scale factor that we can identify as "today": we are at 13.8 billion years since the beginning, just 4.5 billion years since dark energy started to dominate the expansion. The fact the scale factor is unity today is simply our choice, which makes it easier for us to relate earlier and later values of the scale factor today, and therefore the relative change of physical scales. It also implies that scales measured today coincide with their comoving values, that is their values when the scale factor has been factored. From Fig. 6.27 we can read off how length scales in the universe change when we go back or forward in time. For example 5.9 billion years ago distances in the universe were just half of their present value; in 6.2 billion years distances will be longer by 50% of their present value (or by a factor of 1.5).

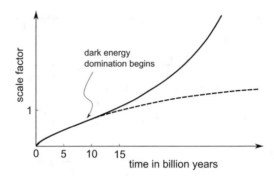

Fig. 6.28 A sketch of the evolution of the scale factor, extending Fig. 6.27 further into the future. The dashed line shows the evolution of the scale factor in the absence of dark energy. In the presence of dark energy, the black full line, we can see the exponential growth of the scale factor and the accelerated expansion beginning when the universe was about 9.3 billion years old

In Fig. 6.28 we sketch how the scale factor will evolve further, extending Fig. 6.27 into the far future, the full black line. On the horizontal axis time in billion years, and the scale factor on the vertical axis. In the presence of dark energy we can clearly see the exponential growth of the scale factor and the accelerated expansion beginning when the universe was about 9.3 billion years old. We also sketch in this figure how the scale factor would have evolved in a universe similar to ours, but without dark energy, shown as a dashed line. In this case the expansion would not have accelerated, but slowed down and eventually coming ever closer to a complete stand still. In this case the scale factor would have continued its gentle growth for ever.

We discussed in detail in Sect. 6.4.2.4 how electromagnetic waves, or radiation, gets redshifted due to the expansion of the universe. When we choose the value of the scale factor to be 1 today, all length scales in the past were smaller by a factor inversely proportional to the redshift (redshift plus 1, this is due to the definition of redshift as the difference between the observed and the emitted wavelength divided by the emitted wave length). How redshift, time, and scale factor are related is also shown in Table 6.2. For example, today the universe is 13.8 billion years old, the redshift is zero and the scale factor is equal to 1 (since we have chosen it to be that value). Another example, at redshift 1, when the universe was 5.9 billion years old, the scale factor had a value 0.5, which means that the universe was just half as big as today and on large scales all distances where just half as large as they are now. For more values see the table.

Table 6.2 A table relating redshift, the time since the beginning of the universe, and the scale factor

Redshift	10^7	100,000	10,000	3400	1100	1000
Time	66 h	75	6900	50,000	380,000	430,000
Scale factor	10^{-7}	10^{-5}	0.0001	0.0003	0.0009	0.001
Redshift	10	4	1	0.31	0.1	0
Time	474 Myrs	1.5 Gyrs	5.9 Gyrs	9.3 Gyrs	12.5 Gyrs	13.8 Gyrs
Scale factor	0.09	0.2	0.5	0.77	0.91	1

Time given in years, if not stated otherwise, "Myrs" stands for million years, "Gyrs" for billion years

Let us now return to the questions with which we started this section, namely what are the age and the size of the universe, since we now have the evolution of the scale factor at our disposal. This allows us to calculate the age of the universe taking into account the dynamic spacetime and the expansion of the universe. We find that the universe is at present 13.8 billion years old. Although we quoted this number already a couple of times, in principle—if we actually would want to do the calculation—we can calculate the age only at this point, having derived how the scale factor and therefore spacetime evolves.

We can now also calculate the size of the universe. As pointed out above, here we mean the size of the visible universe. In a sense the question we need to answer is "how far can we see"?

Let us start by considering a static universe, one that doesn't expand. In this case the answer is straightforward, it is the time light or any other messenger had to travel to an observer now on Earth since the beginning of the universe. As nothing can travel faster than light, and the time that passed since the beginning up to now is simply the age of the universe the answer is 13.8 billion light years. In other words, in a static universe light from a distance of 13.8 billion light years or 4.3 billion parsecs can have reached us by now. Hence the diameter of the visible universe, in the static case, is 8.6 billion parsecs or 27.6 billion lightyears.

However, in the real universe we have taken into account that spacetime is no longer static, the universe is expanding. In this case a light ray or photon, or any other messenger, gets carried with the expansion of space. Does this mean information can travel faster than light? No, since it will also take longer for the information to reach an observer, as there is more space in-between source and observer. We can therefore see farther than in a static universe, and the visible universe is actually 92.8 billion light years or 28.4 billion parsecs in diameter. Light that started its journey 13.8 billion years ago, at distance 46.4 billion light years or 14.2 billion parsecs from us, could have reached us by now, in principle. The actual size of the *visible* universe is bit smaller, as in the first 380,000 years the universe was opaque for all electromagnetic radiation.[40] Nevertheless, it is convenient to use 92.8 billion light years or 28.4 billion parsecs as the size of the observable universe.

Although only information from inside of a sphere 28.4 billion parsecs in diameter can reach us, we shouldn't start thinking of the universe itself as a sphere. We find it more useful to think of the universe as a very large cube or box. If we want, we can take length of the sides of the cube to be 28.4 billion parsecs, such that the observable universe just fits into the box.

It might seem at first counter-intuitive, that we can observe objects more distant or "see" further in an expanding universe, despite objects receding from us. But as discussed above, light from for example a distant galaxy gets carried with the expansion of space. It can travel therefore further in a given time, than in a static universe. But although we can see further, the galaxy in this example we can see is also further away. The net-effect in this case is that we can see the same things as in the non-expanding universe, although they are further away.[41]

The maximum distance we can "see" or receive information from, is referred to as a "horizon", in agreement with the standard usage of the word. We already encountered event horizons earlier in this chapter in Sect. 6.4.3, as the boundary beyond which we can not receive or exchange information from in the case of a black hole. The maximum distance light or information can have travelled to reach an observer since the beginning of the universe is called the "cosmic horizon distance".

The horizon size is growing with time, in a static universe the horizon distance is simply the time since the beginning of the universe multiplied by the speed of light. In an expanding universe the horizon grows faster than in a static universe due to the expansion of space as discussed above. The cosmic horizon will therefore be smaller at earlier times than at late times. By definition, today the cosmic horizon and the visible universe coincide.

The region inside the cosmic horizon is also referred to as a "causal region": inside this region events can in principle influence each other, that is cause an effect. Information can only travel as fast as the speed of light, and for an event to have an effect on an observer, information of the event has to reach the observer. We will return to these concepts in the next chapter.

---

[40]We might however observe gravitational waves from these very early times, as we will briefly discuss in Chap. 9.

[41]Another effect is that the wavelength of the electro-magnetic radiation gets stretched by the expansion. However, this doesn't affect how far we can see directly.

7

How Does the Structure in the Universe Form?

We can now take a closer look at how structure on large scales in the universe forms, having discussed the ingredients for this process in the previous two chapters. These ingredients are the different types of matter and how they behave, and the forces governing the evolution and interaction of the various matter species, in particular gravity.

Matter and galaxies are not randomly distributed in the universe. As we can see for example in Fig. 1.4, a map of galaxies as seen by the 2dF survey, and Fig. 4.12, a similar map made by the SDSS consortium, galaxies are grouped together in structures resembling a web, or a foam.

The "large scale structure" is the distribution of galaxies, clusters of galaxies, and "super-clusters"—clusters of clusters—of galaxies. The processes that lead to this distribution are usually referred to as "structure formation". Since similar processes are also responsible for the distribution of overdense and underdense regions in the early universe, giving rise to what we observe today as the hot and cold spots in the Cosmic Microwave Background, we will also discuss the formation of these structures in this chapter.

7.1 The Final Stages of Gravitational Collapse

Before we turn to structure formation in detail we have to take another look at gravitational collapse, in particular what happens to a cloud consisting of dark matter and baryons. We can safely assume that there is more dark matter than baryonic matter, as observations indicate there is roughly five times the

© Springer Nature Switzerland AG 2019
K. A. Malik, D. R. Matravers, *How Cosmologists Explain the Universe to Friends and Family*, Astronomers' Universe,
https://doi.org/10.1007/978-3-030-32734-7_7

Fig. 7.1 A collapsing cloud consisting of a mixture of dark matter (black and grey) and baryons (blue). On the left we only show the dark matter, at the final stage of its collapse, highlighting some of the dark matter particles. The dark matter particles keep on whizzing about, as they can't shed their kinetic energy, and can't collapse further. On the middle panel we show the baryons (blue), at the time when the dark matter can't further collapse (same time as on left panel). The baryons also whizz about, but they can interact with each other and also emit photons (radiation shown as red arrows). The baryons can cool down—loose energy—and therefore collapse further, as shown on the right

amount of dark matter than matter in the form of baryons. The overdensity leads to an increase in spacetime curvature, and it is therefore useful to think of the matter forming its own "gravitational well" from which it cannot escape. Both dark matter and baryons share the same gravitational wells.

Here it is also important to recall, as discussed in Sect. 5.4, that dark matter interacts mainly or exclusively gravitationally,[1] dark matter does not interact with "normal" baryonic matter and electromagnetic radiation—hence the name, dark matter. Dark matter, interacting only through gravity, therefore behaves rather differently than "normal" matter especially in the later stages of gravitational collapse. We are highlighting the differences in Fig. 7.1.

Let's assume that an initially tiny overdensity has collapsed through the gravitational instability into a overdense cloud of dark matter and baryons. This is shown in Fig. 7.1. The situation for the dark matter is shown on the left panel, this is already the final stage of collapse. Although it moves around fairly slowly originally—we assume it is not relativistic—during the collapse the dark matter gains at least some kinetic energy. The dark matter particles therefore whizz around in the gravitational well formed by the original overdensity. However, they have no way to get rid of their kinetic energy, they only (or predominantly) interact gravitationally after all. The dark matter therefore

---

[1]Even if the dark matter would interact through the weak interaction, this interaction is too weak to be relevant here.

remains at the end of the collapse as a diffuse cloud, shown as grey background in Fig. 7.1 (in all three panels).

The situation is different for the baryons. They start out similarly to the dark matter, as shown in the middle panel of Fig. 7.1. However, the baryons can interact with themselves, and also interact with—or couple to—photons. That means they can emit radiation and cool down, unlike the dark matter. This allows the baryons to radiate away some of their kinetic energy, to form denser structures as discussed in Sect. 6.4.3 in the previous chapter. They will cool further and the baryonic cloud will break up into smaller clouds that further collapse, or they form baryonic objects like stars directly, depending on the size of the cloud and the epoch the collapse takes place. The difference in behaviour between baryons and dark matter when collapsing is also the reason why we did not have to consider the dark matter when discussing the collapse to form "small scale objects" like stars in Sect. 6.4.3 above.

We therefore end up with dense baryonic structures forming inside of rather diffuse clouds, or halos, of dark matter. This is in agreement with numerical simulations, and the simulated universes also agree well with observations. We should stress here that we can only directly observe luminous baryons, stars and very hot gas, and not the non-luminous baryons and the dark matter.[2] We have to keep this in mind when we compare theoretical predictions to observations.

7.2 How to Start: Initial Conditions

In order to understand how structures on large scales in the universe form and evolve, we also need to know how this process started. How were matter and radiation distributed at the beginning of this evolution? We can rephrase this question and ask what the initial conditions were for structure formation. Since we haven't so far explained properly what we actually mean by "initial conditions", let us introduce this concept in a little more detail, before we finally discuss how the large scale structure formed.

We discussed the forces that shape our universe in the previous chapter. It is however not enough to know how the constituents of the universe evolve. We also need to know how, where and when, this evolution started. The description of the "how, where and when" is usually summed up as setting

---

[2]As mentioned above, there is five times as much dark matter as there are baryons, and of the baryons only about 10% are luminous, that is in form of stars or extremely hot gas.

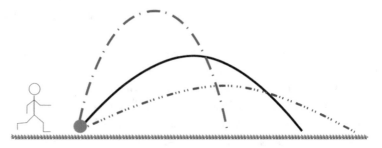

Fig. 7.2 Initial conditions. The ball will follow different trajectories depending on the initial position, speed and the initial direction—the initial conditions in this example

the "initial conditions". This is equally true for the evolution of the universe, as it is for a very simple, mundane example that we give now.

In Fig. 7.2 we show how initial conditions affect a simple physical setup, the kicking of a ball. The trajectory of the ball is not only determined by the forces acting on the ball, but also by the initial conditions which in this example are: the kick-off point, in space and time, the speed with which the ball is launched and the direction it has initially.[3] Or to put it another way, the forces acting on the ball determine its possible trajectories. But from these infinite possible or allowed (by physical law) trajectories, the initial conditions pick out a single, unique one.

As stressed before, usually when we try to solve a physical problem, be it the trajectory of a ball or the evolution of the universe, we would rewrite the problem in terms of mathematical equations. These governing equations also require us to specify initial conditions, that allow us to pick from an infinite number of possible solutions the unique one for the specific problem.

In the example sketched in Fig. 7.2, we highlight the role played by initial conditions in the simple case of kicking a ball. The same holds true for the evolution of the universe. The laws of physics, as discussed in Chap. 6, determine all possible solutions, albeit in this case much more complicated than the simple trajectories in the example. Nevertheless, the initial conditions then pick a particular solution, a particular set of trajectories for all particles in the universe.

The initial conditions determine in principle the evolution of the system, in our case the universe, completely. However, this is taking a "classical physics" viewpoint. Quantum mechanics introduces randomness to the evolution. We

---

[3]This is a simplified example, we neglect air resistance, the fact that the ball is elastic and also not a point particle, and many other effects.

should point out that also "classical" systems, ignoring quantum effects, can show unpredictable behaviour if their evolution is non-linear. As discussed in Sect. 6.4, gravity is non-linear, therefore even infinitesimally small changes in the initial conditions can lead to dramatic changes in the evolution and destiny of the universe. Figure 7.2 shows however a simple, linear system, where the initial conditions indeed determine the further evolution completely.

In Chap. 9 below, we will discuss the physics involved in how the initial conditions for the further evolution of the universe are set during inflation, and how quantum mechanics also has a hand in providing these initial conditions in an elegant way.

7.3   The Formation of Large Scale Structure

We now have all the ingredients to explain in more detail the formation of structure in the universe. By large scale structure we mean the distribution of galaxies, clusters of galaxies, clusters of clusters of galaxies, or super-clusters. We discussed in Sect. 4.4 galaxy clusters, such as the Virgo super-cluster, shown in Fig. 4.11, to which our own group of galaxies belongs. The largest super-clusters, such as the Laniakea super-cluster, also discussed in Sect. 4.4, can contain up to about 100,000 galaxies. But the term "large scale structure" extends beyond just super-clusters. For example when we discussed Figs. 1.4 and 4.12, the maps of the large scale structure made by the 2dF and SDSS galaxy surveys, we already pointed out the web-like structure of the galaxy distribution on large scales. The galaxies are not only arranged in groups but also in filaments, surrounding large, roughly spherical voids spanning hundreds of millions of lightyears. This gives the large scale structure also a foam-like or spongy appearance. However, cosmologist usually speak about the cosmic web, and not about the cosmic sponge.

The formation of structure on large scales started strictly speaking immediately after the end of inflation, about 10^{-34} s after the beginning. Inflation itself was too rapid to allow structures to form, it did however set the initial conditions for the further evolution, as described in the previous section.

At the heart of structure formation is the gravitational instability which we discussed in Sect. 6.4.3 above. The process is complicated, made more difficult by the non-linear nature of gravity, involving a large number of length scales and time scales. We therefore try to highlight what is going on with the help of Fig. 7.3, a sketch of structure formation. We show a cross section through the same region of space, using instead of physical scales and coordinates, comoving scales and coordinates, that is factoring out the expansion of the

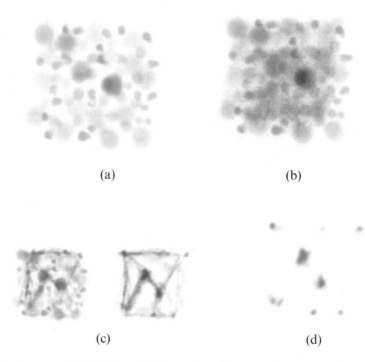

(a)   (b)

(c)   (d)

Fig. 7.3 A cartoon highlighting the formation of large scale structure, from the end of inflation at a tiny fraction of a second after the beginning to the forming of the first baryonic structures, roughly a hundred million years after the beginning. Darker regions indicate higher density. (**a**) The universe at the end of inflation. Matter and radiation are smoothly distributed, with tiny density fluctuations of the order of 1 part in 100,000, formed during inflation. (**b**) Dark matter falls into gravitational wells of the tiny density fluctuations, enhancing and amplifying the primordial density fluctuations generated during inflation. (**c**) After decoupling: further, non-linear evolution, over- and underdensities become more pronounced, now also baryons collapse. Baryons highlighted in red in the figure on the right. (**d**) The same as the previous figure, but only showing baryon overdensities

universe, for presentation purposes. At the end of inflation, roughly 10^{-34} s after the beginning, scales compared to today were a factor of about 10^{-28} smaller. This would make figures using physical scales, instead of comoving scales, extremely impractical.

The initial conditions, determining the further evolution of the universe and the formation of the large scale structure, are set during inflation as highlighted in the previous section. From Sect. 6.4 we are familiar with the concept of mass or energy bending spacetime. We can again use the "rubber-sheet" analogy of spacetime and try and visualise how the energy density of a small overdensity forms a small "dent" in spacetime—a gravitational potential well—, whereas a larger overdensity corresponds to a deeper potential well. Also in the last chapter we defined overdensities as regions that are slightly denser than the average universe, and underdensities as regions slightly less dense than the average (underdensities are only useful if measured relative to the non-zero average density, since densities can't be smaller than zero—we can't have negative mass). Finally, with these definitions in mind, the initial conditions for structure formation in terms of over- and underdensities are small, of the order of just 1 part in 100,000 larger or smaller than the average density.

In Fig. 7.3a we sketch the over- and underdense regions, initially formed during inflation, darker regions indicate a higher matter density. From the end of inflation dark matter can "fall" into these wells, thereby increasing their "depth" and gravitational pull. Already underdense regions get further depleted of material, their matter getting attracted by the overdense regions and flowing towards them. Between the end of inflation and the time of decoupling it is nearly exclusively the dark matter that can collapse and enhance the potential wells, because the universe is still hot enough for the photons to interact with the baryons, providing the baryons with sufficient pressure to resist the gravitational pull.[4] Therefore only the dark matter overdensities grow, overdensities in the photon-baryon fluid do not, they oscillate. The growth and enhancement of the initial, primordial overdensities by dark matter falling into the primordial potential wells is shown in Fig. 7.3b.

Only when the universe has cooled down sufficiently due to its expansion, the photons can decouple from the baryons and the baryons then behave like pressureless particles (on fairly large scales) and start to collapse, falling into the initial potential wells, by this time, 380,000 years after the beginning, already enhanced by the infall of dark matter. The dark matter therefore gives structure formation a "head start". This process is shown in Fig. 7.3c, on the left at early times after decoupling the dark matter structure becomes more pronounced as dark matter forms filaments that take on matter from regions that become underdense and later on empty voids. On the right of Fig. 7.3c we show the same region as in all previous figures several tens of million years after the beginning. In Fig. 7.3d we finally show only the baryons, inside the dark matter overdensities, but without the dark matter. These baryons then collapse further to form very dense structures such as stars. As discussed in Sect. 7.1, because the baryons can interact with photons, they can radiate away energy— they cool down—which would otherwise prevent them from forming dense

---

[4]The Jeans length of the photon-baryon fluid is too large for the fluid to be able to collapse.

structures and objects. The dark matter does not interact with photons and can therefore not loose energy through radiating, it keeps whizzing around, although at a fairly small speeds compared to the speed of light. The dark matter in the galactic halos therefore remains a diffuse cloud or halo, and can never form denser structures; the same holds for the cosmic web in general.

The baryons on the other hand can settle down in the gravitational wells after decoupling. It took roughly 100 million years, or until about redshift 30, until the first stars formed inside rather small dark matter clouds or halos. The first population of stars is called "Population III", and these stars are much heavier than the later generation of stars, roughly one hundred solar masses. Their lifetime is much shorter than later generation of stars and these Population III stars explode as supernovae after tens of millions to hundreds of millions of years, as we discussed in Sect. 6.4.3 above. In these explosions the heavy elements, such as carbon and oxygen, are formed and distributed across the universe. The stars form inside of structures of the order 10^6 or a million solar masses. There are therefore only hundreds of stars in these first structures, and cosmologists usually don't even refer to them as galaxies but use the term "mini-haloes".

The next generation of stars, with masses in the range of about 0.5–10 solar masses, then forms from the debris of the earliest stars and the leftover primordial gas. Over the next several hundred million years these mini-haloes and the stars that they house merge and form the first galaxies, by about redshift 10, or about 500 million years after the beginning. The first galaxies had masses of the order of 10^8 or a hundred million solar masses.

Mergers continue up to the present time, but are far less frequent, resulting in galaxy masses of the order of 10^{12} solar masses, and of the order of 10^{11}, or hundreds of billion stars. Besides merging, galaxies gain further mass by "collecting" or "hoovering up" matter from their surroundings through the gravitational instability.

Our present understanding of the process of structure formation is that smaller structures form early on, grow, and then merge to form larger structures, that similarly keep growing and merge to form larger and larger structures. This process continues to the present day and is usually referred to as "hierarchical structure formation".

But what stops baryonic structures from collapsing altogether, what prevents stars in star clusters and galaxies, or galaxies in galaxy clusters from merging? Recall our brief discussion of energy conservation in Sect. 6.4.2.1, in particular Fig. 6.9: the total energy of a system, its potential and its kinetic energy, is conserved. Let us take a closer look at the collapse of a structure, such as a cluster of stars or galaxies. To gain some intuition for the underlying physical processes, we can model the collective gravitational well of the cluster by a shallow bowl, and the stars or galaxies by some marbles,[5] similar to Fig. 6.9.

We place a couple of marbles at various points at the inside of the bowl and release them. At the beginning of the "formation process" or collapse the marbles only have potential energy, or if we also give them a little push on their way, very little kinetic energy (their speeds are tiny initially). But then the marbles begin to move towards the centre of the bowl and pick up speed, they loose potential energy but gain kinetic energy. Because the marbles start at different positions, they will arrive at different times at the bottom of the bowl in the centre, and simply move through the centre and up again on the other side. Because of little imperfections in the bowl and the initial little shoves we gave the marbles they will not just move in straight lines or trajectories but in circular or elliptical "orbits".

The same happens when a cluster of stars or galaxies forms. Instead of all the stars or galaxies ending up at the centre of the potential well—at the bottom of the bowl—the stars or galaxies keep moving around in the potential well. Because there are no friction forces acting on them, the stars and galaxies cannot get rid of their kinetic energy and keep on moving around. This is a major difference to the simple bowl-marble analogy, the marbles experience friction and therefore eventually come to rest at the centre of the bowl. But in a real collapsing structure, such as some stars or galaxies in their collective gravitational well, will at the beginning of the formation process each have some potential and possibly a little kinetic energy. However, their total energy is dominated by the potential energy, as at the beginning they only move very slowly. The galaxies will experience each others gravitational attraction, and as a result move towards the centre of the collective potential well, loosing potential energy but gaining kinetic energy in exchange, as the total energy is conserved.

The kinetic energy gained by the galaxies increases their velocities which leaves them moving around in orbits in the potential well. They do not all collapse into the centre of the system, they have no possibility to get rid of or dissipate away their kinetic energy and hence keep moving around on their orbits in the gravitational well. The orbits they are moving on are more complicated than the simple ones the marbles follow. This is because the stars and galaxies do not only experience the collective potential well, they also attract each other individually gravitationally. Unlike for the marbles, the

---

[5]We should stress that is a much simplified model, the actual gravitational well is three dimensional, whereas the bowl is two dimensional—recall the rubber sheet analogy in Sect. 6.4.2.2.

likelihood that two stars crash into each other head-on is vanishingly small, compared to the size of the potential well stars are tiny.

This description of processes on astronomical scales is similar to the discussion of the different behaviour of normal and dark matter in the final stages of gravitational collapse at the beginning of this chapter in Sect. 7.1, on smaller scales. However, the stars and the galaxies take here the place of the dark matter particles. On cosmological very small scales, gas clouds can collapse by radiating away the kinetic energy of the gas molecules or atoms on the time scale of a few tens of thousands of years. Although stars do radiate away energy, this radiation has nothing to do with the kinetic energy of the stars. The amount of kinetic energy is also too large, and it would take much longer than the age of the universe for the stars loosing energy at this rate to slow down and eventually collapse.

The stars and galaxies can only interact gravitationally, whereas individual baryons can interact through electromagnetic force, which is much stronger than gravity as discussed in Sect. 6.1. Therefore the baryons in a gas cloud can emit electromagnetic radiation, loose energy and cool down to form stars and planets.

The distances involved also "work against" stars and galaxies. Stars are typically separated by parsecs, if they are for example in the disc of a spiral galaxy, like the sun in the Milky Way. They can be of the order of a couple of hundred times closer in denser regions like the centre of the galaxy, but this keeps the distances still vast. The sun has a diameter 1.4 million kilometres, whereas a parsec is roughly 31 thousand billion kilometres. So even at the centre of our galaxy stars are separated by hundred billion kilometres. Therefore close encounters are fairly rare, and collisions extremely seldom.

Only once very massive objects like neutron stars get very close and form a bound system, they can generate gravitational waves and can loose energy efficiently that way. We discussed this in Sect. 6.4.2.3 for the case of two merging black holes.

Galaxies are typically separated by roughly a million parsec. If they encounter each other, or even collide, the stars in the galaxies again do not collide, and the galaxies can pass through each other at the first encounter. Eventually the galaxies will however merge, and the galaxies change their shape in the collision.

We mentioned above and in previous chapters "gravitationally bound objects", without properly defining what we mean by the term. Objects are gravitationally bound if they "live in" or occupy a collective potential well. They are therefore gravitationally attracted to each other, having lost some potential energy and gained some kinetic energy in exchange. Hence they move around, either in a rather chaotic fashion, as in for example a cluster of stars or galaxies, or fairly orderly as in a spiral galaxy.

We can also ask what limits the size of gravitationally bound objects, such as super-clusters or clusters of clusters of galaxies. In Sect. 6.4.4 we discussed the size of the observable universe and defined the term "causal region". Inside of such a region events can in principle influence each other. It seems like a fair assumption that in order for objects to gravitationally bind they should be able to gravitationally interact, that is influence each other. Hence there is an upper limit for the size of gravitationally bound structures—the size of a causal region. Two regions that are separated by more than the "causal distance", the spatial extent of the causal region, can not influence each other and therefore not collapse into a single bound object. The causal distance is sometimes also referred to as the "Hubble scale", because it is directly proportional to the inverse of the Hubble parameter.[6]

But forming structure also takes time, it is a process and not an instantaneous event. Hence the actual gravitationally bound structures that we observe today are much smaller than a causal region, as matter (normal, baryonic and dark matter) travels not at the speed of light and can collapse only at much smaller speeds. As we saw above in the example of marbles in the bowl, forming a gravitationally bound structures takes time. The causal region is roughly 14 billion parsec in size, but the largest observed structures today are "only" of the order of 100 million parsec. This is still very large, but 140 times smaller than the causal region today.

As discussed above, in hierarchical structure formation small structures form first, grow and merge to form larger and larger structures. Does this mean that this will continue into the distant future? No, if the universe keeps being dark energy dominated this process is coming to an end soon, as the accelerated expansion of space eventually suppresses the formation of new gravitationally bound structures. Structure that has already formed keeps however gravitationally bound, as discussed in Sect. 6.4.2.4, the expansion of space doesn't affect these structures.

In summary, inflation provides the initial condition, the primordial distribution of gravitational wells. The dark matter then amplifies these primordial structures further through the gravitational instability, and provides the "nurseries" for the baryons where galaxies and clusters of galaxies can form, through the same mechanism. Since the baryons can loose energy by emitting radiation, that is they can cool down, dense baryonic structures, such as stars, can form.

---

[6]This can be shown explicitly by studying the governing equations.

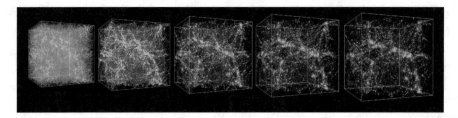

Fig. 7.4 Another computer simulation of the formation of large scale structure. The box size is roughly 130 million lightyears (expanding with universe), starting at redshift 10 or 470 million years after the beginning (on the left), ending today. Only dark matter is shown. *Image credit: simulations performed at the National Center for Supercomputer Applications by Andrey Kravtsov (The University of Chicago) and Anatoly Klypin (New Mexico State University)*

The large scale structure today traces the primordial, initial distribution of over- and underdensities prescribed by the initial conditions. The structure has been enhanced by the gravitational instability. We should stress again, that throughout this process the universe keeps expanding. But as discussed in Sect. 6.4.2.4, structures bound together by gravity "drop out" of the expansion of the universe.

The formation of these large-scale structures is an ongoing process, super-clusters are still forming today, but since dark energy began to dominate the evolution of the universe on the very largest scales "recently"—about 5 billion years ago—the density contrast stopped growing on very large scales. Therefore no structures will form in the future beyond what exists, more or less, already on large scales. However, on scales smaller than a couple of hundred million lightyears, and in particular on small scales, for example inside a galaxy cluster the evolution remains very dynamic. And on for cosmologists very small scales *inside* of galaxies, clouds of hydrogen continue to collapse to form new stars, as we can see for example in Fig. 4.6.

Having sketched and outlined the formation of structure on large scales above and in Fig. 7.3, let us now look at the results of a computer simulation. We already showed the results of a simulation of structure formation in Chap. 2 in Fig. 2.2.[7] In Fig. 7.4 we show the results of another computer simulation of the formation of large scale structure. The size of the box is 43 million parsecs or roughly 130 million lightyears, and the box is expanding with the expansion of the universe—this is also referred to as using "comoving coordinates". We

---

[7]The size of the box in Fig. 2.2 is 430 million lightyears or 140 million parsecs, the results are shown starting at redshift 6 or 650 million years, $z = 2$ or 2.9 billion years, and $z = 0$ today.

therefore do not see that the box itself expands physically by a factor of 10 between the beginning and the end of the simulation. The simulations starts at redshift 10 or 470 million years after the beginning on the left of the figure, ending on the right today; only dark matter is shown. We see how from small overdensities structure forms through the gravitational instability, forming filaments that span the length of the box today. We also see voids forming, in this simulation measuring half the size of the box.

The results of the simulations are in good agreement with the observed and measured distribution of galaxies on very large scales, as shown in Figs. 1.4 and 4.12, for example. When comparing the figures with the simulations we should however bare in mind, that in these figures only a slice of, or a cut through the universe is shown.

A further complication in studying structure formation is that what we can see and observe, galaxies, and the main matter component, namely dark matter, are different. We can only see luminous matter, that is baryons emitting radiation. However, we already discussed that there is about five times as much dark matter as there is baryonic matter. Worse, most of the baryonic matter is in the form of *cold* gas—and hence not emitting radiation—, and not in form of stars that shine. Only about 10% of the baryonic matter has collapsed to form stars, the other roughly 90% is therefore also not visible and directly observable. Hydrogen gas clouds at the centre of galaxy clusters can get heated up to such high temperatures that they emit radiation and therefore can be observed. The reason why the gas is still very hot, of the order of several thousand kelvin mechanism is at moment not well understood. It has been suggested that the gas is couldn't cool down effectively from its high primordial temperatures and is further heated by compression, or that there are additional heating sources; the heating mechanism could be due to quasar-like objects emitting radiation and thereby heating the gas, although this is at the moment unclear. Also occasionally cold gas clouds have been observed when they obscured brighter objects behind them—"in silhouette". But to observe these dark clouds depends very much on a "lucky" alignment of the dark cloud and the bright objects behind them, in order for them to be observed by astronomers.

Therefore what dominates the evolution of the large scale structure, the dark, non-luminous matter, and what we can observe, the luminous matter in form of stars residing inside galaxies, are different, complicating our discussion. When cosmologists calculate the evolution of structure, using for example numerical simulations as shown in Fig. 7.4, the dark matter dominates the evolution to such an extent that ignoring the baryonic matter doesn't change the results much, which simplifies the simulations. However, in observations the luminous matter is all we can see, at least directly. Cosmologist assume however, that the luminous matter, the galaxies, is tracing the underlying dark matter distribution. What we observe in the distribution of galaxies in the universe today is therefore a highly "processed" picture of the universe at the end of inflation.

7.4   The Formation of Structure in the First 380,000 Years

Let us now return to the early universe. Although no baryonic structures can form during the first 380,000 years in the history of the universe, the underlying dark matter over- and underdensities do lead to observable effects. The pattern of hot and cold spots in the maps of the Cosmic Microwave Background, for example taken by the Planck satellite and shown in Figs. 1.3 and 4.14, is made through the interplay of the radiation and the dark matter distribution through the gravitational instability.

We discussed in the previous chapter, that during the first 380,000 years the universe was extremely hot, eventually cooling down to roughly 3000 K at the time of decoupling, due to the expansion of the universe. During these early times the temperature is so high that the baryonic matter present, mainly hydrogen and helium, exists in form of a plasma: the electrons can not combine with the nuclei, as explained in Sect. 5.1.2. The free electrons can however interact with the photons, and although they do not combine with the protons to form hydrogen, they also interact with protons, that is the hydrogen nuclei, and the helium nuclei.[8] The mixture forms a tightly coupled photon-baryon fluid, as the photons are "coupled" to the electrons, and the electrons interact also with the protons and helium nuclei. We will discuss the physics of decoupling further in Sect. 8.2.3. The photon density is sufficiently high and the photons are energetic enough to provide the photon-baryon fluid with pressure.

Therefore, between the end of inflation and the time of decoupling it is nearly exclusively the dark matter that can collapse and enhance the potential wells formed during inflation, because the universe is still hot enough for the photons to interact with the baryons, providing the photon-baryon fluid with sufficient pressure to resist the gravitational pull, as discussed in Sect. 6.4.3 above.

---

[8] Electrons and protons can interact through the electromagnetic force, as discussed in Sect. 6.2.

What happens when the hot photon-baryon fluid "falls" into the gravitational wells provided and supported by the dark matter? Before decoupling the Jeans length of the photon-baryon fluid is too large for the fluid to collapse and form gravitationally bound structures. This means that the gravity pulls the photon-baryon fluid into the potential wells (due to the underlying dark matter distribution) but instead of collapsing the pressure force then pushes the fluid outward again. The interplay between gravitational pull into the potential wells and the restoring force due to the pressure support of the photon-baryon fluid leads to "acoustic oscillations", or sound waves, in the photon-baryon fluid. This happens on scales roughly equal or smaller to the horizon size at this time. Once the photons decouple from the baryons, they are free to travel the universe. We observe these photons today as the Cosmic Microwave Background, they carry unique information that we can extract today about the "environment", the universe at the time of coupling, when they started their journey. We will return to the Cosmic Microwave Background and how it formed in Sect. 8.2.3 in the next chapter.

There is another observational consequence of the physical processes taking place in the photon-baryon fluid between the end of inflation and decoupling, which does affect the baryons and the distribution of galaxies. Let us consider an overdense region—an overdense, roughly spherical cloud—at the end of inflation, just at the beginning of the radiation dominated epoch, see Fig. 7.5. The relevant matter constituents in the universe at this time are the dark matter and the photon-baryon fluid, and the overdense region is just at a slightly higher density than its surroundings. The dark matter and the photon-baryon fluid occupy the same potential well which is due to the overdensity, but the different behaviour of the dark matter and the photon-baryon fluid are crucial again in the following. The density of the region isn't much higher than the average density, roughly about 1 part in 100,000, but this is enough to "drive" the physical processes we consider in the following.

Let us discuss the photon-baryon fluid first. Because the density in the region is slightly higher than average, the pressure in the overdense region is also higher than in its surroundings. The pressure force in the fluid will then drive a pressure wave from the centre of the overdensity, this is just an acoustic wave in the photon-baryon fluid, behaving in a similar way as a sound wave in air. It is actually the pressure, or the overdensity that travels through the fluid as a spherical shell, the photon-baryon fluid doesn't move.[9] In Fig. 7.5

---

[9]A sound wave in air also doesn't move fluid material, in this case air, with it. Air just gets compressed—gets denser—, and then rarefied, as the wave travels through it.

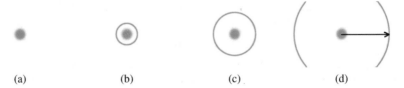

(a)   (b)   (c)   (d)

Fig. 7.5 The generation of baryon acoustic oscillations—a cut through the overdense region. Time increases from **(a)** to **(d)**. On the left—at **(a)**—the small overdensity shortly after inflation ended (overdense dark matter in grey, the overdense photon-baryon fluid in blue). The pressure in the photon-baryon fluid drives a spherical pressure wave outwards, seen here as a blue ring travelling outwards from the centre of the overdensity, whereas the dark matter remains in place—**(b)** and **(c)**. On the right—at **(d)**—after 380,000 years the photons decouple from the baryons, there is no pressure driving the baryons, leaving behind an overdense, spherical shell of "stranded" baryons. The black arrow indicates the distance the pressure wave has travelled in the 380,000 years between the end of inflation and decoupling

the wave is denoted by the blue region. The wave will travel outwards until the photons and the baryons stop interacting and "decouple". At this point in time the photons no longer "drag" the baryons with them, leaving the baryons in the overdense zone stranded at the position they have reached by then (the photons will continue moving outwards). The pressure wave travels with the speed of sound of the photon-baryon fluid, which changes with time but is roughly 60% of the speed of light. The radius of the spherical shell is therefore simply the distance the sound wave travelled between the end of inflation and decoupling, about 145 thousand parsec or roughly 450 thousand lightyears at the time of decoupling.

Meanwhile, the dark matter is not affected by the photons or the baryons, and largely remains at the original position of the cloud (held together by its own gravity). Hence we end up with a small overdensity at the original position, due to the dark matter, and a slight increase in the density in form of a spherical shell at a distance of roughly 145 thousand parsec due to the baryons, at the time of decoupling. Because the universe keeps expanding, 145 thousand parsec at decoupling corresponds to a distance of about 160 million parsec today. The overdense region in the centre of the original overdensity— mainly due to the dark matter—, and the spherical overdense shell—due to the baryons–, make it more likely that galaxies can form at these positions later on, than at other points in space.

In the discussion above we tacitly assumed that the pressure wave can travel outwards unhindered until the time of decoupling, when the pressure support for the baryons provided by the photons ceases. The potential wells due to the overdensities should therefore be rather shallow. What happens to overdensities that are in deeper gravitational wells than the ones described above? In this case the pressure force again pushes a spherical sound wave in the photon-baryon fluid outwards from the centre of the original overdensity. However the gravitational pull of the dark matter at the centre is too strong for the wave to travel outwards until the time of decoupling. The sound wave can only travel outwards for a while, but then—depending on the size and depth of potential well—the overdense shell will travel inwards again. The fluid might even bounce back again a few times before the time of decoupling. These oscillations have given the phenomenon described here its name, "baryon acoustic oscillations" or BAOs.

We therefore end up with sound waves in the photon-baryon fluid oscillating in the deeper potential wells. In this case we also get an overdensity at the centre and a spherical overdense shell due to the acoustic wave or overdensity in the baryon fluid. However, only on a smaller scale than discussed above, that is less than 145 thousand parsec at decoupling (or 160 million parsec today), depending on the position of the wave at the time of decoupling. Since galaxies are more likely to form in the overdense regions, the observational effect is also to increase the probability to find galaxies separated at smaller distances than 160 million parsec today.

Crucially, the maximum distance these pressure waves can travel is 145 thousand parsec at decoupling (or 160 million parsec today). We therefore find that there is a peaked increase in the probability to find galaxies separated 160 million parsec, but not beyond.

Since there are many overdense regions at the end of inflation, these overdense shells overlap, and we do not observe spherical overdensities. But if our assumptions are correct we would expect to find slightly more galaxies separated from each other by a distance of 160 million parsec if we measure the separation of galaxies today, as was done for example in the galaxy surveys discussed in Sect. 4.4. The results of the surveys, maps of the distribution of galaxies, are shown in Figs. 1.4 and 4.12. And this is indeed what the surveys confirmed: a small increase in the likelihood to find galaxies separated by 160 million parsec today, compared to galaxies at a distance of for example 140 or 180 million parsec.

Cosmologists were quite excited when this was observed in 2005. This not only showed that the theoretical predictions agree with the observations. It also gives cosmologist another means to study the composition of the universe: the length scale of the baryon acoustic oscillations today depends on the expansion history of the universe since the time of decoupling, and the expansion history depends on the various contributions to the energy budget of the universe.

Measuring an increase in the likelihood of finding galaxies separated by a distance of 160 million parsec or 480 million lightyears today does not only tell us something about the distribution of galaxies in the universe, it also gives us another means of determining the contributions to the overall energy budget of the basic constituents of the universe. We will revisit the formation of the large scale structure, and cosmic microwave background and baryonic acoustic oscillations, in the next chapter.

8

Going Back in Time

In the previous three chapters we have described the fields and particles that make up the energy content of the universe, and the forces that govern the interaction of these constituents, how the energy content evolves, and how the interplay of the matter and the spacetime geometry leads to the formation of structure. We can use this to put the observations described in the other chapters of this book into a historical context. We choose the point of view of an observer located on Earth, today, and look back out over the evolution of the universe.

Can we go back in time? Not really, but as we have discussed in Sect. 6.4.2.1, the speed of light and that of other messengers, such as gravitational waves, is finite. Therefore when we look out onto the universe we also look back in time! The further an object is away from us, the longer it took information from the object to reach us, and looking further and further we see the evolution of the universe unfolding *backwards*, as shown in Fig. 8.1.

In this chapter we provide a historical overview and highlight important events and epochs in the history of the universe, and where possible tie these to specific observations and the physics discussed so far in the book. Although most of these events and epochs were mentioned in the previous chapters of the book, we can now put them in context and arrange them in their chronological order. We should stress once more that the results presented in the following sections are a synthesis of the physics and observations discussed in the previous chapters. Let us start with a description of the universe today.

© Springer Nature Switzerland AG 2019
K. A. Malik, D. R. Matravers, *How Cosmologists Explain the Universe to Friends and Family*, Astronomers' Universe,
https://doi.org/10.1007/978-3-030-32734-7_8

dark energy begins
to dominate

matter
domination
begins

dark ages

inflation

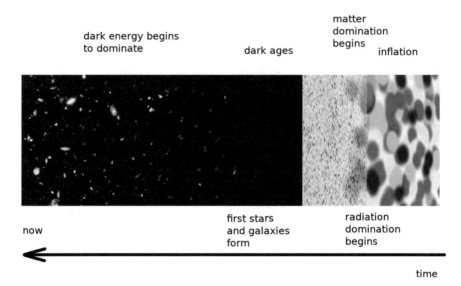

now

first stars
and galaxies
form

radiation
domination
begins

time

Fig. 8.1 The time line of the universe, starting today: "now" is at the left, going back in time towards the right. Particular events discussed in this chapter are highlighted, like the formation of the first stars and galaxies, and the formation of the Cosmic Microwave Background, but also different epochs—dark energy, matter, and radiation domination—and inflation at the beginning

8.1 Now

When is now? We can calculate the age of the universe today by using the expansion history of the universe, how the scale factor changes with time, as discussed in Sect. 6.4.4. By running the expansion backwards, from today to the beginning of the universe we find that the universe 13.8 billion years old.

The age of the universe is of course from a human perspective very large but has otherwise no real, physical significance in the sense that we do not live in a particular, special place in time. We would find the same forces and physical processes at work if we moved a couple of million, or even billion, years back in time or into the future. The first major cosmological change—looking back— took place roughly 6 billion years ago, when the evolution of the universe began to be dominated by dark energy. But before we go that far back in time let us have a closer, and quantitative look at our cosmic surroundings today.

To begin our description of the universe today we make an inventory of the contributions to the energy budget of the universe at present time, which major constituents are there and how large are they. As discussed in Chap. 5, the ingredients that we need are normal or baryonic matter, radiation or photons,

Table 8.1 The matter constituents *today*

Matter species	Radiation	Dark matter	Baryons	Dark energy
Energy density	7.8×10^{-31}	2.2×10^{-27}	4.2×10^{-28}	6.0×10^{-27}
Number density	4×10^{8}	0.1–0.001	0.25	–
Percentage of total	0.01	25	5	69

In the second row the energy densities in kilograms per cubic metre (we give here the *mass* density), in the third row the number of particles per cubic metre (assuming here that dark energy is not a particle), and in the bottom row the contribution in percent to the total energy density (numbers approximate)

dark matter, and dark energy. The values for the number densities are summed up in Table 8.1.

Let us start with the bottom row of Table 8.1, the values already discussed in Chap. 1 and shown in Fig. 1.2 for the energy densities in terms of percentage of the total energy density of the universe. At the present time there is roughly 69% dark energy, 30% matter, of which is 25% dark matter, the rest "normal" baryonic matter, and a tiny fraction of radiation (0.01%). Since 69% of total energy budget is due to the dark energy, the epoch we are living in is dominated by dark energy.

The total energy density of the universe today is 8.7×10^{-27} kg/m^3, and the individual energy densities are for the radiation 7.8×10^{-31}, for the dark matter 2.2×10^{-27}, for the baryonic matter 4.2×10^{-28}, and for the dark energy 6.0×10^{-27}, all in terms of kilogram per cubic metre, see the second row of Table 8.1. We should point out that we actually give the values for the "mass densities" of each constituent, and not the "energy densities". We hope that the mass density is more familiar to the reader than the energy density, which is commonly used in cosmology.[1] These numbers are extremely small if compared to everyday mass densities. For example, water has a density of roughly 1000 kg/m^3, air about 1 kg/m^3 (both under standard conditions). However, the numbers given above and in Table 8.1 are average values, when averaged over all of space. We will return to this point below.

Let us take a closer look at these rather dry numbers and express them in more accessible ways. In the third row of Table 8.1 we converted the energy densities into number densities, that is the number of particles of a particular constituent per cubic metre. In order to do this we need to know the masses of the particles. This easiest for the baryons: the mass of the proton

---

[1]The mass density, or to be precise the rest-mass density, and the energy density are proportional to each other, the constant of proportionality is the square of the speed of light. This is a direct consequence of Einstein's famous mass-energy relation. We follow here the bad practice of many cosmologists and use both terms synonymously.

is 1.673×10^{-27} kg. We choose the proton here, because roughly 75% of the baryons are in form of hydrogen, which has a single proton as its nucleus.[2] We could include the electrons here (although only cosmologists count them as baryons), but this wouldn't change the result of our estimate much, as the mass of the electron at 9.1×10^{-31} kg is much smaller than that of neutrons and protons. The hydrogen atom, consisting of an electron and a single proton as its nucleus, has a mass of 1.674×10^{-27} kg, the difference to the mass of the proton only showing at the third decimal position.

Pinning down the mass of a dark matter particle is trickier, because at the moment the nature of dark matter is still highly speculative, as discussed in Sect. 5.4. However, it is save to assume that the dark matter is non-relativistic and therefore not a light but a rather heavy particle (indeed, that the dark matter is heavy and therefore moving around slowly is necessary to explain the observed structure of the Cosmic Microwave Background, as we discuss below Sect. 8.2.3). The masses for the most popular cold dark matter candidates, WIMPs, range roughly from 1.79×10^{-26} to 1.79×10^{-24} kg per particle.

The rather tiny energy densities translate into small number densities in the case of baryons and dark matter. The number density of the baryons is roughly 0.25 m^{-3}, or 1 baryon per 4 m^3. The number density for the cold dark matter is in the range of 0.1–0.001 m^{-3}. Since it is not even clear whether we can associate a particle with the dark energy, we don't speculate here and leave the entry for the dark energy number density in the third row of Table 8.1 empty.

Let us now turn to the radiation or photons contributing to today's energy budget. The radiation discussed here is the leftover from the early, hot universe, the remnants of which we observe today as the Cosmic Microwave Background radiation. Electromagnetic radiation from stars, "new" photons, contribute only a negligible fraction to the energy density of the radiation leftover from the early universe. As discussed in Sect. 3.3.3, a NASA satellite mission, the "Cosmic Background Explorer" or COBE satellite, was launched in 1989 to study the Cosmic Microwave Background radiation. The results of the mission were published in 1992 and showed that the spectrum of the radiation—the amount of energy radiated at a given wave length—was in agreement with the spectrum emitted by a "black body" with a temperature of 2.73 K. This is what we mean when we say that the temperature of the Cosmic Microwave Background radiation that we receive today is just a little above zero at 2.73 K, or $-270°$ C—when we talk about the radiation in the universe having a

---

[2] Also, the neutron is just a tenth of a percent heavier than the proton, hence our estimate wouldn't change if we had chosen neutrons instead.

particular temperature we mean the temperature of a black body that would emit a similar spectrum.

A black body has the nice property that its electromagnetic spectrum is fully characterised by its temperature: given the temperature we immediately know how much energy the black body emits at a given frequency and wavelength; we already introduced in Sect. 6.4.4 that the energy density of a black body is proportional to the fourth power of the temperature. Knowing the temperature therefore allows us to calculate the energy density, which is how we arrived at the number for the energy density of the radiation given in the second row of Table 8.1. A similar calculation gives us the number density of the photons of a black body, and we find that today there are about 4×10^8, or 400 million photons per cubic metre. We get this very large number density, compared to the other ones in the table, because at low temperatures individual photons have very small energies: photons have no rest-mass and their energy is entirely due to their kinetic energy, which on average at these low temperatures is very small (this is no longer the case when we look at the earlier, hotter universe).[3] There are roughly 800 million photons for every baryon in the universe today.

It is useful to convert some of the above energy densities from the familiar kilograms per cubic metre into more appropriate units for cosmology. We choose a solar mass as mass unit, and to scale up the volume we use parsecs as length unit. The baryon density expressed in these units is roughly 6×10^9 solar masses per cubic million parsec or cubic Megaparsec, that is the baryonic mass in a cube with sides of 1 million parsec or 3 million light years is equivalent to six billion stars similar to our sun. The Milky Way has of the order of 100 billion stars, with varying masses and not all as massive as the sun. But there are also galaxies with fewer stars, and there are stars with less mass than the sun. Furthermore, we should keep in mind that only 10% of the baryonic matter is in form of stars. We therefore expect to find on average one galaxy in a volume of the order of a couple of cubic Megaparsec (for the estimates below, let us choose on average 1 galaxy per 10 cubic Megaparsec). Similar estimates give the typical, average distance between galaxies of being of the order of one million parsec that we quoted in Sect. 4.4, in agreement with observations.

---

[3]To be precise: the photon velocity is always the same (in vacuum), namely the speed of light. The kinetic energy resides in the "speed" or number of oscillations per second, that is the frequency of the electromagnetic radiation. The energy of a photon is directly proportional to the frequency; at low temperatures the photons have on average small energies and low frequencies, at high temperatures large energies and high frequencies.

The reason for the smallness of the energy densities is that although galaxies are extremely big and massive objects, with typical masses in the range of 100 million to 1000 billion solar masses, the universe is an *extremely* big place. Since the densities are simply mass per volume, and although the masses involved are large, the volumes involved are even more mind boggling.

How do we arrive at the values for the energy densities presented in Table 8.1 and in Fig. 1.2? Only the value for the amount of radiation today is a direct measurement, which we get by measuring the temperature of the cosmic microwave background radiation and establishing that the radiation is that of a black body. The other values have to be determined indirectly, because as we discussed in Chap. 5, dark matter and dark energy make their presence felt only indirectly through their gravitational effect, in the case of dark matter, and by affecting the expansion of the universe through its negative pressure, in the case of dark energy. We might hope that it is easier to determine the average density of the baryons, since we can observe them if they happen to be in form of luminous matter. But only about 10% of the baryons are in form of stars, 90% of the baryons do not shine and form gas clouds in galaxies and in the filaments of the cosmic web.

The total energy density is related to the expansion rate as discussed in Sects. 6.4.2.4 and 6.4.4. Therefore by measuring the expansion rate, the Hubble parameter, today we can immediately calculate the present value of the total energy density of the universe. The value of the Hubble parameter today is roughly 70 km/s per million parsec, this means an object one million parsec away will retreat from us at about 70 km/s.

Measuring the Hubble parameter is conceptually straight forward—"all" we have to do is measure the recession speed of galaxies and their distance from us, as discussed in Sects. 4.3 and 6.4.2.4. But in practice this is far from easy, because measuring the distances on these cosmological scales is difficult, as explained in Sect. 4.3. However, once we have measured the Hubble parameter we can calculate the value for the total energy density of the universe today and find the value already given above, 8.7×10^{-27} kg/m^3. Again, converting this into for cosmology more appropriate units we find that the total density of the universe is 1.3×10^{11} solar masses per cubic Megaparsec at the present time. We should keep in mind, that this last result includes all types of matter, baryonic matter, dark matter and dark energy.

The Hubble parameter is not only interesting because it allows us to calculate the total energy density of the universe. It also tells us how fast the universe expands, and therefore gives us information on the dynamics of the universe. The Hubble parameter, the expansion rate of the universe, is defined as the speed or the time rate of change of the scale factor divided by the scale factor, at a given time. The governing equations dictate that the Hubble parameter can only decrease with time, and "at best" remain constant. The speed of the universe's expansion, the time rate of change of the scale factor, on the other hand can decrease, remain constant, or increase. At the moment it is increasing, that is the expansion of space is accelerating. However, for a long period of its history, the expansion of the universe slowed down, it was decelerating. The last time the universe was expanding as fast as it does today was roughly 10 billion years ago. We will discuss when the universe "switched" from decelerating to accelerating in the next section.

The individual energy densities for the dark matter and the dark energy can also only be inferred indirectly. We discussed in Sect. 5.4.1 the observational evidence for dark matter, why the cosmological standard model requires dark matter—galaxy rotation curves and the formation of large scale structure, including the temperature fluctuations in the Cosmic Microwave Background radiation, need the "right" amount of dark matter to be in agreement with the observational data. That we need dark energy in our energy becomes apparent also from studying the Cosmic Microwave Background radiation, from measuring the total energy density today, and from studying the expansion history of the universe, as discussed in Sect. 5.5.1 and detailed below in Sect. 8.2.

We already mentioned above that the moment in the history of the universe we find ourselves in—now—is not particularly remarkable. What can we say about the place we find ourselves in? We described our galactic neighbourhood in Sect. 4.4. The Milky Way, the spiral galaxy that the solar system is part of, is unremarkable beside the fact that it is our cosmic home. As pointed out before, for example in Sect. 6.4.2.4, we are not in a special place. The average values describing the universe today, and therefore also our cosmic environment, as summarised in Table 8.1 are unexceptional and are only remarkable in their smallness and might be taken as further reason for humility.

However, what does make us—and everybody else living in galaxies—special, is that we are at a location of high density. This should not come as a surprise, since we are by necessity on a habitable planet in the solar system, which is part of a galaxy. The densities given in Table 8.1 are *average* values, averaged over the whole universe. Locally these densities can be rather different, and from our discussion in Chap. 7 we expect that galaxies as part of the cosmic web are in regions where the matter density is much higher than on average. We already discussed in Sect. 7.3 that for example all luminous matter is in form of stars in galaxies, and the galaxies are embedded in dark matter halos, and the galactic halos are part of even larger structures made up of dark matter and non-shining baryons, or hydrogen gas, resembling a web on cosmological scales.

To get an idea of how much denser our galactic environment is, let us estimate the dark matter density of the Milky Way, assuming that the dark matter is evenly spread out through the galactic halo—the roughly spherical cloud enveloping the galaxy (see Fig. 5.5). The energy density of dark matter in the Milky Way is then about 2.8×10^{-23} kg/m^3, still a tiny number but nevertheless, this is already roughly 13,000 times, or more than four orders of magnitude, denser than the average dark matter density in the universe.[4]

Although the actual shape and the distribution of the dark matter within the halo of our galaxy are still active areas of research, it is save to assume that the halo has a roughly spherical shape, and the dark matter is smoothly distributed within it. But the dark matter is not spread evenly throughout the halo, it is denser towards the centre and less dense towards the outer regions of the halo. With these more realistic assumptions we can estimate that the dark matter density at the position of the solar system is roughly 100,000 times larger than the average dark matter density in the universe. We would therefore expect to find 100 to 10,000 dark matter particles per cubic metre on earth, depending on the mass of the dark matter particle. As discussed in Sect. 7.1, dark matter doesn't collapse much further since it cannot cool down and shed its kinetic energy (even if the dark matter particles are heavy and their kinetic energy, compared to their rest-mass energy, is small). The dark matter particles keep therefore whizzing around.

Baryonic matter, on the other hand, can collapse to form much denser structures, like stars and planets. The average density of the sun is roughly 1400 kg/m^3, that of earth is 5500 kg/m^3. Comparing this to value in Table 8.1, we therefore see that we live in a place that has a baryonic density that is 1.3×10^{31} times, or 31 orders of magnitude larger than the average in the universe! But this doesn't mean that the Milky Way or earth are special, this holds for all galaxies similar to ours, and all planets similar to ours.

We can also ask, how much of space is taken up by galaxies today in the universe. Using our estimates above that on average there is roughly 1 galaxy per 10 cubic Megaparsec, and take our galaxy as typical,[5] we find that the galaxy takes up only a millionth of this volume. Put another way, if we pick a small galaxy sized volume of space at random, it is very likely we would pick a region

---

[4]We assumed that the halo is a spherical cloud of dark matter, extending out to 250 thousand lightyears, or about 80 thousand parsec from the centre of the Milky Way (five times further than the stars in the galactic disk), and the mass of the Milky Way is 10^{12} or a trillion solar masses. For this rough estimate we can safely neglect that roughly 5% of the mass of the galaxy is in form of baryons.

[5]The Milky Way disc has a diameter of roughly 100 thousand lightyears, hence we choose as "typical" volume a sphere with the same diameter—the sphere that we can fit the Milky Way disc into. The volume of this sphere is roughly 10^{-5} cubic Megaparsec.

of empty space since there are of the order of one million regions without a galaxy compared to the one region that contains a galaxy. Hence being in a galaxy is a bit special.[6]

Let us now briefly discuss what physical processes relevant for cosmology take place in the universe at the present day. As discussed in Chap. 7, structure formation is a process that begins at the end of inflation and continues throughout all epochs in the history of the universe. In particular during matter dominated epoch, the formation of the large scale *visible* structures, the distribution of galaxies and clusters of galaxies, takes place as discussed in Chap. 7. It continues to the present day but slows down in the dark energy dominated epoch. Having explained the physics behind structure formation in the previous chapter, we will highlight in this and the following sections key events relevant for structure formation and cosmology in general, and fit these events into the time-line of the universe.

As discussed in Chap. 7, structure formation takes time and therefore we see the largest structures, super-clusters, that is clusters of cluster of galaxies, becoming gravitationally bound only fairly recently, by cosmological standards a couple of billion years, or even today. By gravitationally bound we mean a system that shares a single potential well and does no longer take part in the expansion of the universe. Although this holds in principle for any system with two or more gravitating bodies, such as the one used as examples in Fig. 6.11, we have here in mind systems containing several galaxies or even galaxy clusters. An example for these structures on the very largest scales is Laniakea super-cluster that we discussed/introduced in Sect. 4.4. This super-cluster contains roughly 100,000 galaxies and is about 150 Mpc or 500 million lightyears large.

At the present day galaxy formation itself has mainly stopped, no new galaxies form as in earlier epochs galaxies forming in overdense regions already used up most of the hydrogen gas in the universe. But galaxies are not static, they evolve as the stars the contain evolve. Galaxies can also merge if they are in close proximity. For example, the Andromeda galaxy (see Fig. 4.2) is today roughly 2.5 million lightyears or 785 thousand parsec away from the Milky Way. Both galaxies are members of the local group and are moving towards each other, colliding in about 4.5 billion years from now. After their collision the two galaxies will merge and loose their distinct spiral shape and possibly form an elliptic galaxy. Since they are in a gravitationally bound system the expansion of the universe will not prevent this collision.

---

[6]Indeed, the volume taken up by a galaxy compared to the space surrounding it is negligible—cosmologists can therefore approximate galaxies frequently in their calculations as "point particles", idealised objects that have a mass but no spatial extension.

Today gravitational collapse takes place also on cosmological very small scales. Stars form through the gravitational collapse of clouds of gas in the present universe, as can be seen for example in Fig. 4.6, the beautiful "Pillars of Creation" in Sect. 4.2.

We end this section with some more observational data characterising our present universe. We saw in Sect. 6.4.4 that the diameter or size of the observable universe today is 28.4 billion parsecs or 92.8 billion light years. A sphere of this size has a volume of 1.2×10^{31} cubic parsec, or 4.2×10^{32} cubic lightyears.

Alternatively, to quell any notions that there is anything spherical about the universe itself other than the region we can observe, we can calculate the volume of a cube with side length 28.4 billion parsecs and arrive at very similar values of 2.3×10^{31} cubic parsec, or 8.0×10^{32} cubic lightyears. Taking our previous estimate of roughly about 1 galaxy every 10 cubic Megaparsec we arrive at a value of roughly 10^{12} or 1000 billion galaxies in the observable or local universe.[7] We should stress again, this is only the observable or local universe. The whole universe is much, much larger. We will return to this topic in the next chapter.

8.2   The Matter and Dark Energy Dominated Epochs

The physical processes relevant for cosmology that take place during early dark energy domination and the late matter dominated epoch—roughly the last 10 billion years—are very similar and we can therefore discuss them together. We already discussed structure formation today in the universe in the previous section, let us next take a closer look at how cosmologists discovered that the universe isn't matter dominated today, and at the time the universe changed from matter to dark energy domination.

---

[7]Using much more sophisticated arguments—taking into account the mass range of stars in galaxies, and that there are different types of galaxies with widely varying numbers of stars—we arrive at similar numbers. But we only want to give here an "order of magnitude correct" result.

8.2.1 The Transition from Matter to Dark Energy Domination

Until fairly recently cosmologists thought that the universe is matter dominated today. However, at the end of the last millennium several observations indicated that the universe is today not matter but dark energy dominated. The evidence came from observations of the Cosmic Microwave Background, or to be precise, observations of the fluctuations in the temperature of the Cosmic Microwave Background, which showed that the geometry of the universe is very simple,[8] and from observations of a particular type of supernovae, so called "Supernovae of type Ia", which showed that the expansion of the universe is not slowing down but accelerating. We will discuss the Cosmic Microwave Background in detail in Sect. 8.2.3 below, and turn now to the supernovae observations.

A supernovae of type Ia is an exploding white dwarf star. As discussed in Sect. 4.3, the sequence of events leading to the explosion of the star for this type of supernova follows the same steps in each case, and hence the brightness, or luminosity, for these explosions is always the same. Supernovae of type Ia are also very bright and therefore visible at large, cosmological distances, making them excellent standard candles. Knowing the absolute brightness of a supernova allows us then to calculate the distance to the supernova from the observed brightness, because the observed brightness of a light source is inversely proportional to the square of the distance to an observer. Another method to determine the distance to the supernova is by measuring its redshift. The light emitted by the supernova gets stretched due to the expansion of space on its way to an observer on earth, as discussed in Sects. 4.3 and 6.4.2.4.

Both methods need to take the expansion of the universe into account, but depend on the expansion in slightly different ways. We have seen in Sect. 6.4.4 that the rate of the expansion, and the expansion itself, depend on the matter content of the universe. Measuring the distances to many supernovae using both methods can therefore give us information on what kind of and how much matter there is via the expansion history of the universe. This is illustrated in Fig. 8.2, where we sketch a brightness versus redshift diagram for type Ia supernovae. The red dots are the observational data, each dot representing a single measurement—the *observed* brightness and redshift of a supernova. Redshift 0.01 corresponds to a distance today of 44.4 million parsec, 0.1–434.3 million parsec, and a redshift of 1–3.4 billion parsec. In

---

[8] We will discuss the geometry of the universe in Sect. 9.3.2.2.

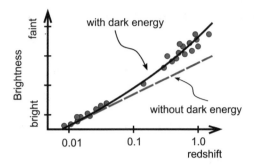

Fig. 8.2 Observational evidence for dark energy: the brightness of type Ia supernovae versus redshift. In a universe without dark energy we would expect to find the data points on the dashed blue line, in a universe containing also dark energy on the black line. Redshift 0.01 corresponds to a distance of 44.4 million parsec today, 0.1–434.3 million parsec, and a redshift of 1–3.4 billion parsec

addition to the observational data we also plot the theoretical predictions of how the brightness of a supernova is related to the redshift. As already mentioned, this relation depends on the expansion of the universe, and hence its matter content. The blue dashed line corresponds to the expansion history of a universe without dark energy, the black line to one including dark energy. We find that the observational data fit the theoretical expansion history for a universe containing dark energy.

This was a unexpected result, as at the time the majority scientific consensus—or prejudice—expected the observations to agree with a matter dominated universe[9]! At low redshifts we see from Fig. 8.2 that the data points agree with both theoretical curves, because differences in the brightness-redshift relation only become apparent at higher redshifts. Until the mid-1990s no precise data at the required redshifts were available, but towards the end of the decade new supernovae observations were made at much greater distances and redshifts, out to redshifts of roughly 1. At these redshifts the difference between the theoretical models becomes much more pronounced. To fit the data, that is to get agreement between the observed brightness-redshift relation and the theoretical predictions, we see from Fig. 8.2 that we need to include dark energy in the total energy budget of the universe. The values for the different contributions that match the observations best are the ones given in

---

[9]Some cosmologists had however even before the supernovae observations were made and analysed entertained the idea that dark energy could play a significant role in the dynamics of the universe, as we will discuss below in Sect. 8.2.3 and in Chap. 9.

Table 8.1, in particular we need to add 69% in terms of dark energy to the total energy budget of the universe.

This result is not only remarkable because we have to add 69% dark energy, another novel, exotic constituent even weirder than dark matter, to the total energy budget of the universe today. The addition of dark energy has another immediate consequence, namely that the expansion of the universe is accelerating! Dark energy gives rise to negative pressure, as discussed in Sect. 5.5, and its physical effect is to speed up the expansion of the universe. This means that the expansion of the universe today is not slowing down but accelerating. As we discussed in Sect. 6.4.4, all other contributions to the energy budget—radiation, baryonic matter and dark matter—lead to a slowing down of the expansion rate, but the supernovae observations lead us to conclude that the speed of expansion is increasing, and hence the expansion is instead accelerating today, requiring us to introduce dark energy as an explanation. The change over from decelerated expansion to accelerated expansion happened roughly 3.8 billion years ago.

The change over from matter domination to dark energy domination is a major event from the point of view of cosmologists. A new constituent in the energy budget is taking over, dark energy, the rate at which the universe is expanding is changing, the expansion switches from deceleration to acceleration. How physical processes proceed during the first couple of billion years of dark energy domination is very similar to these processes taking place in the matter dominated epoch, we can therefore deal with both epochs in this section together. Only in the far future will the expansion have speeded up sufficiently to affect, for example, the collapse and clustering of matter. We will return to this topic in the final chapter, in Sect. 10.2. At present the accelerated expansion is only noticeable on the very largest scales.

At around 10 billion years after the beginning, or roughly 3.8 billion years ago, the matter content—normal matter and dark matter—was diluted to such a level that the dark energy began to dominate the energy content of the universe. The transition from dark energy to matter domination can also be seen in Figs. 6.25 and 6.26 in Sect. 6.4.4. The transition is also referred to as the "time of dark energy and matter equality".

If we would have been around at the time, we wouldn't notice this effect. The turnover took place at a redshift of 0.3, when the size of the universe was about three quarters of its present size, that is the scale factor was roughly 0.75. The Cosmic Microwave Background temperature was 4 K at the time, still very close to today's temperature of roughly 3 K.

Structure formation takes place throughout the history of the universe, beginning at the end of inflation and still ongoing at the present day, as discussed in Chap. 7. The rate of growth the density contrast, the density fluctuations relative to the average density, depends on how fast the universe is expanding.

During dark energy domination the density contrast stops growing on the largest scales, as the expansion of space eventually becomes too rapid even for the gravitational instability to counteract. This will however only happen when the energy content is fully dominated by the dark energy, today it "only" makes up 69% of the total energy as discussed in Chaps. 1 and 5. Also, on smaller scales on the order of a hundred of million parsecs, gravitationally bound structures continue to grow. During the matter dominated epoch the density contrast does grow on all scales. It grows at the same rate, or proportionally, to the scale factor that describes how space on large scales expands. It is in the matter dominated epoch that structure grows the most, compared to the dark energy and radiation dominated epochs. Not only is the growth rate of overdensities during matter domination faster than during radiation domination—they do not grow at all during dark energy domination. But the matter dominated epoch is also the longest epoch in the history of the universe, so far.

8.2.2 The Matter Dominated Epoch Until Decoupling

Going further back in time we have now arrived in the matter dominated epoch, where most of structure formation takes place. We discussed the process of structure formation in some detail in Chap. 7, and here we simply put some of the highlights into the timeline of the universe. As already done previously in this chapter, we start at more recent events working backwards in time.

Galaxies like the Milky Way formed when the universe was roughly 1 billion years old, at a redshift of about 5. As discussed in Sect. 6.4.2.4, the redshift is related to the scale factor which describes how much the expansion of the universe has stretched length scales (at cosmological distances). The scale factor in turn is related to the temperature of the Cosmic Microwave Background[10]; this also shows that the temperature of the Cosmic Microwave Background increases with redshift. A redshift of 5 means that large scales, such as the size of the universe, at the time were only a sixth of today's value, and the

---

[10]The temperature of the Cosmic Microwave Background is inversely proportional to the scale factor, as discussed in Sect. 6.4.4. This also implies that as we go back in time and the scale factor decreases, the temperature increases.

temperature of the Cosmic Microwave Background was six times larger than today, or roughly 18 K.

These "modern" galaxies, of which the Milky Way is a typical example, form through the mergers of galaxies from an earlier, first generation. These first galaxies had much smaller masses, and formed at a redshift of about 10; this process is sketched in Fig. 8.3. Several thousand of these first galaxies had to merge to assemble for example the Milky Way. The Milky Way has a mass of about 1000 billion solar masses, whereas a typical first generation galaxy has a mass of about 100 million solar masses, both values include the more abundant dark matter and the baryonic matter in form of gas and stars. During the merger the gas that had not formed stars yet in these galaxies, and the remnants of exploded stars—leftovers from supernovae—get mixed together, and the gravitational interactions lead to a mixing and churning of the gas and further gravitational collapse. As a result of these interactions we observe that the formation of stars peaks at redshift of about 2, when the universe was roughly 3.3 billion years old or 10.5 billion years ago, more than 2 billion years after modern galaxies like the Milky Way formed. But this only means that the rate at which stars were formed was highest at this time, and lower at other times. New stars nevertheless formed before and after this peak in star formation, for example our own star, the sun, formed "only" 4.6 billion years ago. Stars that formed recently, that is within the last couple of billion years, in rather mature galaxies are referred to as "Population I" stars. These stars continue to form in galaxies today. By the time Population I stars form, the galaxies resemble the ones we can see in our galactic neighbourhood today, and examples are shown in Sect. 4.2.

Going further back in time we see the first generation of galaxies forming at a redshift of about 10, when the universe was roughly 470 million years old, as shown in the middle of Fig. 8.3. At this time large distance scales were roughly an eleventh of their present day values, and the temperature of the Cosmic Microwave Background was about 33 K. As discussed above, these galaxies are much smaller than modern galaxies like the Milky Way and merge during the next several hundred million years. The stars in these first galaxies are called "Population II" stars. These first generation galaxies are at the very edge of what we can observe with the current telescopes. But the galaxies are so far away that their images are just very faint dots. We remind the reader here that far away also means long ago in astronomy, and the light of these first galaxies travelled for more than 13 billion years until it reached observers on earth. The next generation of telescopes, such as the ELT discussed in Sect. 3.4.3, should give us better images of these galaxies.

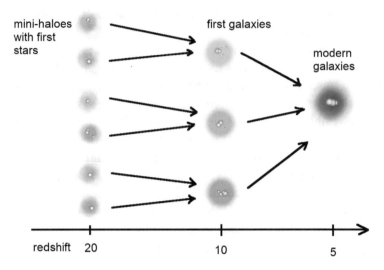

Fig. 8.3 Galaxy formation: today's galaxies formed through the mergers of earlier structures. By a redshift of about 20 the first stars have formed in dark matter "mini-haloes" (shown on the left). These merge to form the first galaxies by a redshift of about 10 (shown in the middle), and these first galaxies merge to form "modern" galaxies like the Milky Way (on the right). Time increases from left to right, redshifts 20, 10, and 5 correspond to roughly 180 million, 470 million, and 1 billion years after the beginning

Continuing our journey back in time we arrive at the formation of the first stars. As discussed in Chap. 7 the early galaxies form from "mini-haloes" or "proto-galaxies" at a redshift of about 20, or 180 million years after the beginning, shown at the left of Fig. 8.3. At this time the temperature has reached about 60 K, and the size of the universe was about 5% of its present size.

The mini-haloes had a mass of about one million solar masses, and contained only a couple of very massive stars. These first stars were of the order of one hundred solar masses and formed from the baryons, mainly in form of hydrogen gas, that accumulate at the centre of the dark matter mini-haloes. It took the gas until this period, when the universe was around 180 million years old, to first cool down sufficiently to form compact, massive objects. These objects were sufficiently massive that when they further contracted under their own gravitational pull, they could support such high pressures and temperatures at their centres that the fusion of hydrogen into helium started. With the start of the fusion reaction these objects turned into the first stars, and this first star generation is called "Population III".

The lifetime of these Population III stars was much shorter than that of the later generations of stars, of the order of tens to a few hundred million years. In comparison, both typical Population I and II stars have a lifetime of the order of 10 billion years, and we therefore find that modern galaxies like the Milky Way contain both Population II and I stars. At the end of their lifetime the Population III stars exploded in supernovae and during these violent explosions the heavy elements, such as carbon and oxygen, were formed.

Shown in Fig. 8.3 and already discussed in Chap. 7, this overall process is known as "hierarchical structure formation": small structures—the mini-haloes—form first, that then merge to form the first galaxies, that then merge to form the familiar modern galaxies, for example the members of the Local Group (see Fig. 4.10 in Sect. 4.4). Only on small scales by cosmological standards, do we observe that structure also forms through the break up of larger objects. For example gas clouds "just" a couple of lightyears across that form a solar-system break up into the star and the surrounding planetary system, as was the case for our solar system, or in the case of the cores of mini-haloes where the regions that host the first stars are of the order of hundreds of lightyears, that break up into several regions that then each collapse to form a massive star. As already discussed, the hierarchical formation of structure has not ended yet, clusters and super-cluster of galaxies continue to form to this day.

But let us return to the universe at redshift 20. Hydrogen is the most common baryonic matter, and the hydrogen gas present at this time is in form of "neutral hydrogen". This means that in the hydrogen atom the nucleus, a single proton, and its electron are bound together by the electromagnetic force. However, as soon as the first stars begin to shine the situation changes through the radiation emitted by the stars. The light from these first stars started to "kick" out the electron from the hydrogen atom and separate the electron from the proton, that is ionising the neutral hydrogen that was not yet in form of dense objects and stars. At the beginning of this ionisation process there are only few stars in the universe, and the radiation from these stars at first ionised the regions close to the stars. But as time passes the radiation reaches further regions and also more stars form until eventually all the free hydrogen in the universe is ionised.

This ionisation process begins with the formation of the first stars at redshift 20 and by about redshift 5, when the universe was 1 billion years old, all the free hydrogen has been ionised. By studying when and how the ionisation process happened we can learn a lot about the universe at this epoch. For example when the first stars formed, how many there were, and what kind of stars.

We encountered ionised material, or plasma, in previous chapters and will discuss in detail in Sect. 8.2.3 that the baryons in the universe were already ionised before this epoch, in the early universe before a redshift of about 1100, or 380,000 years after the beginning. All the baryonic matter—mostly hydrogen—was in form of a plasma because the photons that today make up the Cosmic Microwave background were energetic enough to keep the electrons from combining with the protons. This changed at redshift 1100 during "decoupling" when the universe had cooled down enough for the photons to be no longer energetic enough to interact with the electrons and keeping them from binding to the protons. The protons could then combine with the electrons to form neutral hydrogen and the universe remained neutral until the first stars began to shine and their radiation started to ionise hydrogen in the universe again. Since the universe was ionised before this second ionisation process this is usually called "re-ionisation".

In our journey back in time we have arrived at the period named the "dark ages" by cosmologists. The name has been chosen because there are no sources of light in the universe at this time, as stars have not formed yet. But this does not mean that nothing interesting or relevant for cosmology is happening during this period, the very opposite is the case. Throughout this period structure formation continues, and it is during this epoch that the mini-haloes discussed above assemble: the baryons—mostly neutral hydrogen—that got caught in the gravitational potential wells of the mini-haloes can begin to collapse[11] and sink to the centre of the wells to form the first baryonic structures. It is these structures, clouds mainly consisting of hydrogen gas, that cool down and through the gravitational instability collapse and get more and more dense to form the first stars that then also end the cosmic dark ages.

The cosmic dark ages begin after decoupling and recombination when the universe was 380,000 years old, at a redshift of 1100, and last until the first stars begin to shine at a redshift of about 20, about 180 million years later. The temperature in the Cosmic Microwave Background radiation falls from about 3000 K at the beginning of the dark ages to 60 K at the end. This point in the history of the universe is also referred to as "cosmic dawn".

The universe continues to be filled with the radiation that we today observe as Cosmic Microwave Background, that cools down from about 3000 K at decoupling and the beginning of the dark ages, bathing the universe in infrared radiation and bright orange light (the black body radiation peaks at a wave

---

[11]After decoupling the photons do not interact with electrons any longer and the electrons can combine with the free protons to form neutral hydrogen. But as discussed above, there is therefore no more photon-baryon fluid and the photons can also no longer provide pressure to the baryons. The baryons can collapse.

length of about 10^{-6} m, or 1 µm, which is the near infrared). But this radiation, or these photons that started their journey at 380,000 years after the beginning gets redshifted, and crucially, there are no stars or any other objects yet that would be sources of *new* radiation, until the end of the dark ages.

A lot of research is focused today on the study of this epoch in the history of the universe. For example the next generation of radio telescopes, such as the SKA discussed in Sect. 3.4.1, might be able to pick up faint emissions of the neutral hydrogen, which emits radiation at wave lengths of 21 cm. Hydrogen atoms emit this 21 cm radiation spontaneously, but this happens extremely rarely, roughly once every 10 million years for one neutral hydrogen atom.

To make up for the rarity of this emission, there are luckily enough hydrogen atoms around to use 21 cm radiation to map the neutral hydrogen: let us pick as an example redshift 100, when the universe was 16.5 million years old. At this time we find from Table 8.1 that there are roughly 250,000 baryons per cubic metre (the baryon number density grows with the third power of the redshift— going back in time), and most baryons are in form of hydrogen, hence baryon number is a good estimate for the number of hydrogen atoms. A single hydrogen atom spontaneously emits 21 cm radiation once every 10 million years, or 3×10^{-15}/s. This does not look very promising, as it is a very small number, even if we have 250,000 hydrogen atoms. However, if we instead of considering a volume of just one cubic metre consider scales and volumes more appropriate for cosmology, and take the number of hydrogen atoms in a cubic parsec,[12] we find a much, much larger number, namely roughly 10^{40} transitions or spontaneous emissions per second. This makes it possible to use these tiny and extremely rare emissions of 21 cm radiation from each hydrogen atom to be collectively observed by astronomers today over distances of billions of parsecs. Mapping the distribution of the neutral hydrogen will allow cosmologists to tap into a new and vast source of information about the universe *before* the first stars and galaxies formed.

We should therefore not think about the cosmic dark ages as a period that is completely obscured from observations and not really relevant for the history of the universe—the very opposite is the case. During the cosmic dark ages many essential processes that shape the further evolution of the universe take place, they are simply not emitting light or radiation that is in the visible range of the electromagnetic spectrum. However, if we study the dark ages in the history of the universe using the 21 cm emissions of the neutral hydrogen that was present at the time, this epoch becomes a rich source of information

---

[12]At redshift 100 there are about 7×10^{49} baryons in a volume of 1 cubic parsec.

for cosmologists today. The aim is to use the 21 cm radiation to map the distribution of the neutral hydrogen in the universe after decoupling, just as we can use the visible light emissions of galaxies to map the large scale structure in the later universe as discussed in Sect. 4.4. The 21 cm emissions are another source of information about the universe which promises to be even richer than the Cosmic Microwave Background (that we will discuss in the next section). The reason is that we receive the Cosmic Microwave Radiation only from a single point in time, the time of decoupling, whereas we observe the 21 cm emissions from a whole epoch, the dark ages. Therefore we observe the Cosmic Microwave Background as a two-dimensional surface, whereas the 21 cm maps are three-dimensional.

Into the dark ages falls also the epoch in the history of the universe when the temperature of the Cosmic Microwave Background was "pleasant" by human standards. At a redshift of 109 the temperature of the universe was 25 °C, when the universe was 14.6 million years old. As far as we know, there is no deep significance associated with these values and there is therefore no "pleasant epoch" in the history of the universe that attracts the attention of cosmologists at the moment. Let us therefore now return to the official history of the universe and a particularly important event for cosmologists, the formation of the Cosmic Microwave Background.

8.2.3 Cosmic Microwave Background Formation

The Cosmic Microwave Background has made an appearance in most chapters of this book so far, and we already touched on its formation in Sect. 7.4, in the context of the formation of structure on large scales in the universe. We have now arrived at the point in the history of the universe where the electromagnetic radiation that we observe as the Cosmic Microwave Background began its journey, and we therefore discuss its formation in more detail. The Cosmic Microwave Background is one of the richest sources of information about the early universe, which makes it such an exciting topic for cosmologists.

We discussed in Sect. 7.4 that the Cosmic Microwave Background, like all structure in the universe, started to form after inflation, but what we observe today is a "snapshot" of the universe at redshift 1100, when the universe was 380,000 years old, and the formation process ended. The size of the universe was by a factor of 1/1100 smaller than today. The temperature of the Cosmic Microwave Background radiation was 3000 K at this time, and the composition of the universe was very different compared to today, as we already saw in Fig. 1.2 (on the left): the universe contained 63% dark matter,

24% radiation, and 12% of baryonic or normal matter, as percentage of the overall energy budget. Compared with today, there is a considerable amount of radiation present at this time, but the universe is "still"—we are moving back in time—matter dominated. Dark energy plays no role in the energy budget at this early time.

What we observe today as the Cosmic Microwave Background radiation are the photons that were until the end of the formation process coupled to the baryons. Let us therefore begin with a closer look at the physics of decoupling itself. The term "decoupling" is both used for the physical process and the time in the history of the universe when this process happened.

The time at which decoupling takes place is determined by how the universe expands, or its expansion history. As discussed in Sect. 6.4.4 as the universe expands, the temperature of the electromagnetic radiation that fills it decreases—the temperature of the radiation is inversely proportional to the scale factor which describes how physical distances on large scales evolve. As the radiation cools the photons that make up the radiation become less energetic— we usually describe this by the photons cooling down. Decoupling takes place when the photons are sufficiently cool to no longer keep the electrons from combining with the protons to form hydrogen nuclei, and this happens when the temperature is about 3000 K and the universe is about 380,000 years old.

Before the time of decoupling the temperature in the universe is so high that the baryonic matter is in the form of a plasma, the electrons are separated from the nuclei, which are just protons in the case hydrogen (the most abundant element[13]). The photons and electrons interact with each other through the electromagnetic force, they are "coupled together", see Fig. 8.4 on the left. Here it is useful to think about the electromagnetic radiation in terms of particles, that is in terms of photons. In the figure protons are red spheres, electrons are grey spheres, and the photons are black dots (the black lines are the photon paths). The density of photons and electrons is very high, hence a photon bumps into an electron and scatters off it before bumping into the next electron, this process repeating itself frequently as shown in the figure on the left. The photon-baryon fluid filling the universe is opaque, like a fog, while the photons bounce off the electrons before decoupling.

The negatively charged electrons also interact with the positively charged protons through the electromagnetic force. Hence the baryons, to which we

---

[13] At the time of decoupling there are also helium nuclei, consisting of two protons and two neutrons, but they only make up a quarter of the baryonic mass and will also be ionised, so we can ignore them in our discussion.

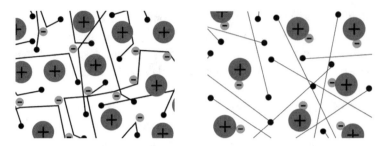

Fig. 8.4 Radiation and matter—electrons and protons—before and after decoupling (red spheres are protons, grey spheres electrons, black dots photons). On the left: before decoupling the universe is very hot, the photons are energetic enough to keep the electrons from binding to the protons; the photons can interact with the free electrons, and the negatively charged electrons interact with positively charged protons. On the right after decoupling: the universe has cooled down enough for the electrons to combine with the protons to form hydrogen; the photons no longer scatter off electrons and can travel freely

here count electrons and protons, together with the photons form the photon-baryon fluid we encountered and discussed above in Sect. 7.4.

After decoupling the photon-baryon fluid-"fog" that fills the universe at the time becomes transparent to the photons, because the electrons can now combine with the protons to form neutral hydrogen, and there is nothing for the photons to interact with, as shown on the right of Fig. 8.4. After a very short period there are no free electrons left that could interact with the photons. We should keep in mind that decoupling happens everywhere, at roughly the same time in the universe, and all the photons are then free to travel from where ever they happen to be at decoupling. In which direction the photons travel in is random, they travel in all directions.

The photons can then travel unhindered from the time of decoupling until today—if they don't hit an obstacle on their journey and get absorbed. We can observe these photons today, if they happen to travel in our direction and if they started their journey at the "right distance" to arrive today. Since the photons are from the time of decoupling they some times are referred to as the oldest light in the universe that we can observe. At the start of their journey they would have constituted a dark red glow, but their wavelengths get stretched by the expansion of the universe on their way to us—they get redshifted (see Fig. 6.18). The wavelengths with the highest intensity are no longer in the near-infrared range and are instead in the microwave range. We therefore can't observe the Cosmic Microwave Background radiation with optical telescopes today, only with microwave telescopes.

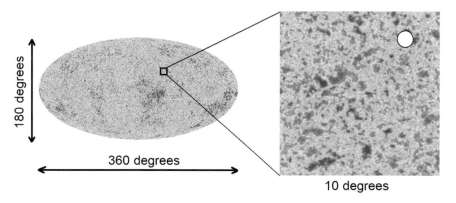

Fig. 8.5 The Cosmic Microwave Background: on the left all of the surface of last scattering, mapped onto an ellipse, as in Fig. 1.3. On the right a "zoom" into a 10° by 10° section of the map. The circle in the top right hand corner has a diameter of 1°, roughly the size of the horizon at the time of decoupling. Red corresponds to hotter, blue to colder than average temperature (at the level of 1 part in a 100,000). Temperature fluctuations on scales larger than 1° are due to photons "escaping" potential wells generated before decoupling, fluctuations on smaller than 1° are due to acoustic oscillations of the photon-baryon fluid just before decoupling

The universe is today filled everywhere with this radiation, and was at the time of decoupling. However in order for us to observe the photons from the time of decoupling, they must arrive today. That means that the photons have the same travel time and therefore travelled the same distance. Hence what we observe today has the appearance of the inside of a sphere,[14] the diameter of which is nearly as big as the particle horizon today. The sphere is slightly smaller than the particle horizon, because the Cosmic Microwave Background photons started their journey 380,000 after the beginning, "missing" the distance that photons could have travelled in these early years. Since decoupling also implies that this is the time the photon scattered for the last time off the electrons, the inside of this sphere is also known as the "last scattering surface".

Figure 1.3 shows an all sky map of the Cosmic Microwave Background, that is the surface of last scattering, which is also shown on the left of Fig. 8.5. For any observer the Cosmic Microwave Background has the appearance of a sphere, with a radius of roughly 14 billion parsec or 45 billion lightyears.[15] The

---

[14]Because decoupling did not take place exactly instantaneously, we observe the Cosmic Microwave Background photons that originate from a thin shell, with a thickness today of about 50 million parsec. This is however tiny compared to its diameter and we can neglect this here.

[15]The distance to the horizon, the size of the visible universe is slightly larger, roughly 14.2 billion parsec or 46 billion lightyears, as discussed in Sect. 6.4.4.

surface of this sphere, or to be precise the interior surface, is then projected onto an ellipse, a process similar to carefully peeling a satsuma and then stretching the peel into elliptical shape, as discussed in Chap. 1. The ellipse in Fig. 8.5 is therefore the projection of all of the microwave sky, as can also be seen from the angles. The reader might be familiar with the geographic coordinate system on earth: specifying two angles, one for "longitude" in the range from 0 to 360°—equivalent to moving along, or parallel to the equator—and one for "latitude" in the range of 0 to 180° allows us to identify any point on the surface, and therefore similarly on a map of this surface. This also holds for the surface of last scattering, the only differences being that we here have the inside of the "celestial sphere", and the sphere is much, much larger.

Let us now turn to the pattern of density fluctuations that we observe in the Cosmic Microwave Background radiation today as tiny deviations from its average temperature, at the level of 1 part in 100,000, as shown in Figs. 1.3 and 8.5. As already touched upon in Sect. 7.4 the Cosmic Microwave Background formed through the gravitational instability, in the same way that the large scale structure, the distribution of galaxies in the present day universe, did. In the case of the large scale structure we observe the electromagnetic radiation emitted by stars in galaxies in the "late" universe, whereas in the case of the Cosmic Microwave Background we observe the radiation that was coupled to the baryons for the first 380,000 years in the history of the universe and were "trapped" in the potential wells of the dark matter until decoupling.

In previous chapters we saw that it is often convenient not to discuss concrete numbers but to keep the discussion general and focus on relations and relative sizes, because the relevant quantities change with time. For example, we saw in Sect. 6.4.3 that for a particular region to collapse and form a gravitationally bound object, the region needs to be larger than a particular size or length scale, which we introduced as the Jeans length. But the Jeans length depends on the composition of the universe, which changes with time, and therefore also the Jeans length changes with time. We saw in Sect. 7.3 that the largest region that can collapse at a given time is a "causal region", a region smaller than the causal horizon size at this time. Again, this size depends—through the Hubble parameter—on time. These two length scales govern structure formation, and therefore also control the formation of the Cosmic Microwave Background. However, since we are interested here in a particular time, the time just before decoupling, we can give below also the numerical values for the physical sizes of these length scales.

At the time of decoupling the energy density of the dark matter dominates the energy budget, as there is 63% dark matter but only 24% radiation and

12% baryonic matter (as shown in Fig. 1.2). The baryon-photon fluid will therefore "follow" the underlying dark matter distribution and "fall" into the potential wells due to the dark matter.[16] How the fluid then behaves depends on the size of the potential wells. Material or fluid in potential wells larger than the horizon scale cannot collapse, but material in potential wells or regions smaller than the horizon scale—regions that are "inside the horizon"—*can* collapse. If the region is inside of the horizon and larger than the Jeans length it will collapse and the fluid or material form a gravitationally bound object; if the region is inside the horizon but smaller than the Jeans length, the fluid will begin to collapse and be attracted to the centre of the region but bounce back, leading to the fluid oscillating in the potential well, without forming a gravitationally bound object.

The causal or particle horizon had a diameter of 250 thousand parsec at decoupling, this scale has been stretched by the expansion of the universe to roughly 275 million parsec at the present day. Only regions of horizon size or smaller *can* collapse. But whether a region then collapses to form a gravitationally bound, dense object depends on the size of the region and its matter content, as discussed in Sect. 6.4.3. If the region is smaller than the Jeans length, the matter will only begin to collapse and will then bounce back, driven back by the pressure forces. At this time, just before decoupling, the Jeans length measured 660 thousand parsec, today 726 million parsec. The baryon-photon fluid in potential wells smaller than the horizon therefore didn't collapse to form bound objects, but oscillated in the potential wells (because the Jeans length was larger than these structures at the time). In this case the horizon distance, or scale, therefore separates regions that neither collapse nor oscillate, and regions where the baryon-photon plasma oscillates in the potential wells (if the size of the potential well fits inside the horizon).

The sound horizon is the furthest distance a sound wave can travel during a certain time, in this case from the end of inflation and the beginning of the radiation dominated era to the time of decoupling, roughly 380,000 years after the beginning. It is similar to the particle horizon, the furthest distance a particle or photon can travel, but instead of travelling at the speed of light, a sound wave travels at the speed of sound. The sound speed of the photon-baryon fluid is roughly 60% of the speed of light at this time. We discussed the sound horizon already in Sect. 7.4, and at decoupling this scale measures 145 thousand parsec, which gets stretched by the expansion to 160 million

---

[16] This is a slight simplification as also the photons and the baryons contribute to the potential wells—all forms of energy curve spacetime. But at decoupling there is nearly twice as much energy in the dark matter, than in the other components.

parsec today. Let us recall, that sound waves are patterns of compression and rarefaction, that is oscillations in the pressure of a fluid. These pressure changes lead also to density changes in the fluid. The size of these fluctuations is however tiny, of the order of just 1 part in 100,000, or 10^{-5}, of the average density value.

The photon-baryon fluid in a potential well of about the size of the sound horizon had just enough time to start collapsing and get maximally compressed, before bouncing back by the time of decoupling. Since compression means higher density, and therefore also higher temperature, this will result in a region hotter than the average temperature, a "hot spot" in the Cosmic Microwave Background map, of the order of the sound horizon in size. There are many of these sound horizon sized hot regions, as many as there are potential wells of this size. It took the photon-baryon fluid in the sound horizon sized potential wells 380,000 years, the age of the universe at this time, to undergo half a period of these oscillations.[17]

But there are also smaller regions—smaller potential wells—, where the fluid had enough time to get compressed and bounce back, and even smaller potential wells were several cycles of compression and rarefaction took place. However, in each case compression will increase the density and temperature of the fluid, rarefaction will decrease both. What we end up with are hot and cold regions in the baryon-photon fluid.

An important feature of the underlying pattern that forms is that all oscillations started at the same time, and therefore oscillations of a given period or wavelength are in phase. For example, all oscillations in potential wells of roughly the sound horizon size will be at maximum compression at the time of decoupling, and have therefore also the same temperature. The same then holds for smaller potential wells. This leads to the distinctive pattern of hot and cold spots in the Cosmic Microwave Background.

At decoupling, the oscillations of the baryon-photon fluid stop, because the photons can no longer provide pressure support for the baryons: the baryons begin to collapse (the Jeans length for the baryons decreases because the photons no longer interact with the baryons, and there is nearly no pressure), and the photons begin their journey through the universe. This is why we can say the Cosmic Microwave Background is a "snapshot" of the universe at the time of decoupling.

---

[17] A full oscillation cycle of compression and bounce back or rarefaction for oscillations of sound horizon size would take two times 380,000 years.

Above we discussed the "fate" of the baryon-photon fluid inside the potential wells provided by the dark matter. Let us now turn to the photons, and consider the time just after decoupling, when the photons are no longer "tied" to the baryons and can start their journey through the universe, until some of them get observed by us today.

We can distinguish the behaviour of the photons depending on the size of the potential well they find themselves in at the time of decoupling. Photons in potential wells that are larger than the particle horizon at the time of decoupling have to climb out of these wells and loose energy in the process, they get redshifted and cool down. For photons in potential wells smaller than the horizon we find two effects: firstly, if they are in an underdense region, they are cooler than average, if they are in an overdense region they are hotter. The second effect is that the photons will again loose energy when they climb out of the smaller potential wells, but this effect is usually smaller than the change in temperature due to compression and rarefaction. The photons therefore carry the information about their environment at decoupling with them, until some of these photons reach us today.

When describing the density and temperature fluctuations of the Cosmic Microwave Background we often use angles, instead of distances. This is because we see the Cosmic Microwave Background as a spherical surface, the surface of last scattering, we can use two angles to specify any point on this surface—we don't need to specify a third coordinate, since we assume the surface of last scattering to be two-dimensional (its thickness is usually negligible, and therefore the distance to the surface is the same everywhere). On the largest scales the universe is isotropic, the same in all directions in direction.[18] The tiny density fluctuation, and the equivalent fluctuations in the temperature, can therefore also be described as deviations from isotropy, and are therefore also referred to as "anisotropies" when discussing the Cosmic Microwave Background.

These deviations from the average temperature, the temperature fluctuations or anisotropies are of the order of 1 in 100,000, and have particular angular sizes, corresponding to the size of the hot and cold regions at decoupling.

As discussed above, the causal or particle horizon at decoupling has a diameter today of roughly 275 million parsec. This distance corresponds to

---

[18]We remind the reader of the Copernican principle discussed in Sect. 6.4.2.4, that the universe on the largest scales is the same in all directions and the same in all places—the universe is isotropic and homogeneous.

an angular scale of 1.1° on the surface of last scattering for any observer today, see Fig. 8.5 (this is the angle the horizon distance subtends on the surface of last scattering, when viewed today from Earth). Similarly the sound horizon measuring 160 million parsec today, corresponds on the surface of last scattering to an angle subtending 0.7°.

We therefore see in Fig. 8.5 hot and cold regions, or hot and cold "spots" on the surface of last scattering, of angular size of less than 0.7°, stemming from the oscillating regions. We also see temperature fluctuations of angular size larger than 1°, which are due to photons climbing out of potential wells in the case of overdense regions formed during inflation.

The photons of the Cosmic Microwave Background carry information about the very early universe from the time when it was just 380,000 old and the surface of last scattering formed. By analysing the distribution of hot and cold spots cosmologists can learn a lot about the universe at the time of decoupling, but also about earlier times.

Structures, that is hot and cold spots, with angular sizes larger than 1° on the surface of last scattering map out the potential wells and the distribution of the dark matter that formed during inflation. Photons trapped until decoupling in these "super-horizon" or larger than the horizon potential wells will loose energy and cool down when they climb out of their potential wells and are released to travel across the universe at decoupling. They will lose less energy and cool less if the potential well is more shallow, hence this region will appear in comparison hotter. We therefore gain information on large scales about the size and the depth of the potential wells that were generated during inflation.

Because there can be no oscillations on scales larger than the horizon, and the sound horizon is the largest distance that sound waves can have travelled, we can identify the size of the sound horizon with the largest scale on which we observe oscillations in the photon-baryon fluid. The size of the sound horizon depends on the sound speed of the fluid, and since the sound speed depends on how much radiation relative to the baryon density there is, we can calculate the photon to baryon ratio at the time of decoupling.

The amplitude of the oscillations depends on the depth of the potential wells, and therefore on the amount of dark matter present at the time of decoupling and before. The Cosmic Microwave Background therefore also provides us with another independent piece of evidence for the existence of dark matter.

The restoring force for the oscillations is the pressure of the photon-baryon fluid—as discussed previously, after the fluid "falls" into the potential wells it bounces back due to the pressure force. The size of the pressure force depends on the amount of photons present at the time, but the re-bounce will also depend on the combined amount of photons and baryons (more baryons means a larger mass that has to bounce back, hence the response will be more sluggish, the photons have to "push" more mass around when the fluid bounces).

By measuring the distribution of hot and cold spots on the surface of last scattering we also get information about the underlying geometry of the universe on the very largest scales, scales on which the universe appears homogeneous and isotropic. The path that the photons take on their journey through the universe depends on this geometry, and because of this also the angular size of the hot and cold spots that we observe depends on the geometry. The measurements are in agreement with a flat geometry. We will discuss the geometry of the universe further in Sects. 9.3.1.2 and 9.3.2.2.

The Cosmic Microwave Background therefore provides us with a unique window on the early universe at the time of decoupling, just 380,000 years after the beginning. By studying the distribution of hot and cold spots on the surface of last scattering we can test our models of the early and late universe and study the processes that lead to the formation of the Cosmic Microwave Background.

8.3 Radiation Domination

In our journey back in time we have now reached the radiation dominated epoch, lasting from the end of inflation to the beginning of the matter dominated epoch discussed in the previous section. In this epoch, the radiation determines the evolution of the universe.

As already done previously in this chapter, we shall start at more recent events working backwards in time. The radiation dominated epoch begins with matter-radiation equality which is defined as the time in the history of the universe when the energy per volume in matter and in radiation was equal. The transition from matter to radiation domination can also be seen in Fig. 6.25 in Sect. 6.4.4, and is due to the different responses of the radiation and the matter to the expansion of the universe. The transition took place when the universe was 50,000 years old, and the temperature was roughly 9000 K, at a redshift of about 3300. The size of the universe was smaller by a factor of 1/3300.

The energy density in the radiation was larger by a factor of 1.2×10^{14} or 14 orders of magnitude compared to today, and the matter energy density was larger by a factor of 3.6×10^{10} or 10 orders of magnitude. The values for the energy densities of matter and radiation are the same at this time—by definition—, roughly 9×10^{-17} kg/m^3.

Matter-radiation equality has no direct, immediate physical consequences, however the expansion rate of the universe changes at this point in the history. It is also interesting to note, that at this point the universe has heated up so much that the density of electromagnetic radiation, something we usually don't think of as being "dense", has become as large as the density of the normal and dark matter combined. Indeed, going further back in time there is more energy in the radiation than in all the other components taken together.

During radiation domination the density contrast[19] grows on very large scales—scales larger than the sound horizon—but slower than later on during matter domination. On smaller scales density fluctuation will bounce back due to radiation pressure. No stable baryonic structure can therefore form, as the baryons are prevented from collapsing, they are coupled to the photons (recall, decoupling takes place during matter domination, 330,000 years *after* matter-radiation equality).

As discussed in the previous section, before decoupling the normal or baryonic matter is in the form of a plasma. Therefore, at matter-radiation equality the universe is filled by the photon-baryon fluid—the temperature is even higher than at decoupling and too high to allow the electrons to bind to the nuclei.

As we continue our journey back in time the temperature increases further, until when the universe was about 20 min old we have reached the end of nucleosynthesis, when the universe was by a factor of about 10^{-8} smaller and the temperature was about 100 million or 10^8 K. We discussed nucleosynthesis, or "Big-Bang Nucleosynthesis" as it is also referred to, in Sect. 5.2. During this period the nuclei of the light elements—helium, lithium, and beryllium—formed, but only in tiny amounts besides helium, which makes up 24% of the baryonic matter. Hydrogen nuclei, protons, make up rest or 75% of the baryonic matter. At the start of nucleosynthesis, when the universe was just 2 min old, the temperature was about one billion or 10^9 K.

During the first 2 min of the universe the temperature rises further and the universe is so hot that the heavier nuclei that form during nucleosynthesis are still in the form of their constituent particles, protons and neutrons. But as we continue back in time, at about 1 s after the beginning, the temperature is so high that finally also the neutrons and protons, and any other particles that

---

[19]As discussed previously, the density contrast describes the fluctuations of the density relative to the average density of the universe.

consist of quarks, will give up their identities and the universe is filled with a quark-gluon soup.[20] At this time the temperature has reached about ten billion or 10^{10} K, and the size of the universe is by a factor 10^{-10} smaller than today.

In the previous chapters we have not mentioned neutrinos much. Due to their tiny masses they play only a minor role in the evolution of the universe, from a large scale structure point of view. Neutrinos play an important role in nuclear processes involving the weak force, for example in the fusion reaction in the cores of stars and in supernova explosions. But because the conditions in the very early universe are very similar to the cores of stars—high temperatures, densities and pressure—neutrinos do play a role at early times.

Despite interacting with other particles only through the weak interaction,[21] at temperatures of above roughly 10^{10} or 10 billion kelvin they nevertheless get coupled to the other particles and the radiation, since the densities involved are so high. Similarly to the decoupling of the baryons from the photons, we also have "neutrino-decoupling", the time after (and temperature below) which neutrinos stop interacting with the other constituents. Neutrino decoupling took place when the universe was roughly 1 s old, and the size was less than a billionth of today's size.

Neutrinos are extremely numerous, there are today roughly 3×10^8, or 300 million neutrinos per cubic metre in the universe. However, since they interact so weakly we don't notice being bathed in neutrinos. At the time of neutrino-decoupling, there were of the order of 10^{38} neutrinos per cubic metre!

The neutrino background might also become 1 day another source of information about the universe, the neutrinos carrying information from the time when the universe was just 1 s old, roughly 380,000 years earlier than when the Cosmic Microwave Background formed. Observing the neutrinos is however very difficult as they interact only weakly with normal matter, as discussed in Sect. 3.3.4.1 on non-standard messengers.

As we go further back in time the temperature keeps rising, and even more exotic processes can take place. However, these processes also involve more complicated particle physics and we won't discuss them here further. When the universe was about 10^{-12} s old, the temperature in the universe was roughly 10^{18} K, and the size of the universe by a factor of 10^{-18} smaller than today. Some cosmologists use this time as the beginning of the radiation dominated epoch, since this is the time when we can apply standard particle physics to describe the contents of the universe.

---

[20] Particles like protons and neutrons are made up "quarks", held together by "gluons", which mediate the strong force.

[21] Neutrinos also interact through gravity, but this is not relevant here.

Finally, we arrive at the epoch of inflation, which we shall discuss in detail in the next chapter. Inflation ends at roughly 10^{-34} s after the beginning. Between the end of inflation and the time after which "standard" particle physics applies, at 10^{-12} s, the field or fields driving inflation decay into the normal matter fields and radiation that fill the universe since. We take the end of inflation as the beginning of the radiation dominated epoch.

The time at the end of inflation, and inflation itself, is less well understood than the previous epochs. Therefore we should take most numbers involving inflation as order of magnitude estimates. We will discuss inflation in the next chapter.

In Table 8.2 below we sum up some of the key events in the history of the universe as discussed in this book, the time they took place, and the temperature at which they took place. The times and temperatures are approximate values.

Table 8.2 A summary of key events in the history of the universe, and the temperature at the time

Time	Key event	Temperature (K)	Comment
"0"	"The Beginning"	"infinity"	Unknown
10^{-36} s	Inflation	10^{27} K	The earliest time cosmologists can "agree" on
10^{-34} s	Radiation domination	10^{18} K	Universe dominated by radiation, standard particles subdominant
1 s	Neutrino decoupling	10^{10} K	Neutrinos decouple and begin their journey
100 s	Nucleosynthesis	10^8 K	Lightest nuclei form
50,000 years	Equality	9500 K	Universe filled with equal amounts of matter and radiation
380,000 years	Decoupling	3000 K	CMB photons begin their journey
180 million years	End of dark ages	60	Proto-galaxies form, first stars "shine"
500 million years	First galaxies form	33	Proto-galaxies merge
1 billion years	Modern galaxies form	18	First galaxies merge
10 billion years	Dark energy takes over	4 K	Late time acceleration begins
13.8 billion years	Now	2.73 K	Today

Both times and temperatures are approximate values

9

How Did It All Begin?

Our journey through time in the previous chapter has taken us to the very earliest epoch in the history of the universe that cosmologists are still confident to talk about and, more importantly, should also be listened to. This epoch is called inflation.

9.1 What Is Inflation?

Let us begin with describing and defining inflation. Inflation started at about 10^{-36} s after the beginning and lasted until about 10^{-34} s which is extremely early in the history of the universe, and a very short period of time.[1] However, during this very short period the universe expands by a factor of 10^{26}, or 26 orders of magnitude.[2] This is an extremely large amount of expansion in a very short period of time, which can only be achieved if the universe grows exponentially, and the speed of the expansion increases with time—the expansion of the universe is accelerating. We already encountered an epoch of accelerated expansion in the late universe, when the dark energy begins to dominate the evolution although the expansion is much less rapid than during inflation. We can therefore define inflation as a short period of rapid, accelerated expansion in the very early universe.

---

[1] We should keep in mind that 10^{-36} is a rather tiny number, there are 35 zeroes after the decimal point before the 1 appears.

[2] Some models of inflation predict an even larger amount of expansion.

© Springer Nature Switzerland AG 2019
K. A. Malik, D. R. Matravers, *How Cosmologists Explain the Universe to Friends and Family*, Astronomers' Universe,
https://doi.org/10.1007/978-3-030-32734-7_9

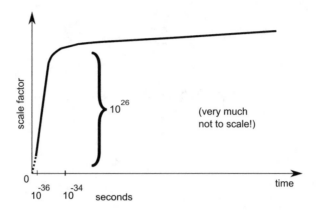

Fig. 9.1 The evolution of the scale factor during inflation. During an extremely short period of time very early on in its history, the size of the universe, as described by the scale factor, increases by a factor of 10^{26}, or 26 orders of magnitude. The figure is not to scale

The expansion of the universe during inflation, as described by the scale factor, is illustrated in Fig. 9.1. This extremely rapid expansion takes place before the "normal" evolution of the scale factor shown in Fig. 6.27 in Sect. 6.4.4 (Fig. 6.27 is the continuation of Fig. 9.1). After its growth by 26 orders of magnitude during inflation, the scale factor keeps growing slowly for another 18 orders of magnitude or a by a factor of 10^{18} afterwards. Let us emphasise: during the evolution of the universe from the end of inflation until today the universe grows in size by 18 orders of magnitude taking 13.8 billion years, during inflation it grows by 26 orders of magnitude taking the tiniest fraction of a second.

Let us illustrate these numbers with some examples. The size of an atom is typically about 10^{-10} m, expanded by a factor of 10^{26} the atom would have a size of 10^{16} m, or roughly 1 lightyear.[3] If we expand the atom further by another 18 orders of magnitude, the atom would be expanded to 10^{18} lightyears or a billion billion lightyears. This huge number dwarfs even the size of the visible universe today which is roughly 93 billion lightyears in diameter!

This immediately suggests the question, how big was the visible universe at the beginning of inflation? Let us first move to the end of inflation, when the size of visible universe today was just 9.3×10^{-8} lightyears in diameter, which is in SI units roughly 9×10^8 m, or 900,000 km. But at the beginning of

---

[3] We should keep in mind, however, that no atoms exist during inflation.

inflation the visible universe was a mere 9×10^{-18} m, smaller than the diameter of a proton which measures about 10^{-15} m!

9.2   What Drives Inflation?

In order to drive inflation, or to be precise, to get accelerated expansion of the universe, we need negative pressure. This is—again—a concept that can be much easier understood and seen to arise from the governing equations. Negative pressure doesn't exist in Newtonian physics, where we only have *relative* negative pressure, for example when water gets sucked up through a pipe.

In general relativity we can also have *absolute* negative pressure. Without using the governing equations explicitly, let us use a particular form of dark energy—a cosmological constant—to get an idea of how negative pressure comes about.

A cosmological constant has the same value, which corresponds to the same energy density, throughout the universe and at all times. The equation that describes the evolution of the energy density is the "energy conservation equation". From the equation we find that for the energy density to be constant, and not diluted by the expansion of the universe as is the case for matter and radiation (see Sect. 6.4.4), we need to introduce a quantity that counteracts the dilution. The only quantity in the equation that can counteract the dilution is a negative pressure. Therefore for a particular energy density not to be diluted by the expansion of the universe, it must give rise to negative pressure. This is the case for the cosmological constant, but also for other forms of dark energy, such as "scalar fields".

How can scalar fields give rise to negative pressure? A scalar field has the property that for a period of time the field can change very, very slowly. The scalar field changes so slowly with time and is the same *everywhere* in the universe, that it mimics the cosmological constant during this period and also gives rise to negative pressure, accelerating the expansion of the universe. After a while the scalar field changes more quickly, it stops providing negative pressure, and the universe stops its accelerated expansion. This is exactly what cosmologists think happened during inflation: the scalar field remained changing very slowly for a short period of time—during inflation—and then started to change more quickly, thereby ending inflation. After inflation the scalar field decays into radiation, the standard matter fields, and dark matter.

During inflation the energy density in the scalar field remains roughly constant, similar to the energy density of the cosmological constant. This implies that the Hubble parameter and therefore also the size of the horizon remain roughly constant during inflation.

The physical effect of negative pressure is to counteract the gravitational slow down of the expansion due to the matter in the universe: baryonic matter, radiation, and dark matter all gravitate leading to the gravitational instability discussed in Sect. 6.4.3. Even (normal positive) pressure acts as a source for gravity, and therefore all these types of matter slow down the expansion of the universe. Only negative pressure can accelerate the expansion, and counteract the slow down of an expanding universe, or counteract the possible collapse of a non-expanding universe.

The cosmological constant was originally introduced in 1917 by Einstein for this purpose, to counteract the collapse of his static universe model. But he dropped the concept when Hubble discovered that the universe is expanding, which dynamically kept the universe from collapsing.

Why do we introduce scalar fields instead of a cosmological constant, if both lead to negative pressure and an accelerated expansion? We can't have something constant driving inflation, because we need inflation to end! Otherwise no structures would have been able to form in the universe, and there would be no matter fields. We need the source of the negative pressure to stop driving the accelerated expansion at some point in time, therefore it can't be a constant.

It is for this reason that cosmologists assume that the rapid acceleration is driven by one or more scalar fields.[4] The epoch of inflation is dominated by scalar fields and the field driving inflation is called the "inflaton". If there are other scalar fields present they play a subdominant role (often referred to as "spectator-fields"). However, one scalar field is all that is needed for inflation. All scalar fields, including the inflaton, decay at the end of inflation into radiation, normal matter and dark matter.

But why did cosmologists introduce such a weird epoch of accelerated expansion in the early universe in the first place? We will discuss the reasons in the next section.

---

[4] The number of scalar fields depends on the particular model of inflation. Different models of inflation have different physical and observational consequences, like the amount of accelerated expansion they provide and the distribution of density fluctuations they generate, see Sect. 9.4.

9.3 Cosmological Puzzles Explained by Inflation

In the 1980s some cosmologists noted that the accepted model of the universe and its history that they were using at the time left a few things unexplained. For instance theorists noted that astronomers do not see any magnetic monopoles which then popular models of particle physics predicted should be present in large numbers.

Also no explanation was available of how it has come about that the universe is isotropic and almost homogeneous, looking the same in all directions and being the same everywhere, despite the fact that many regions of the universe have never been in contact during the lifetime of the universe. For instance, how could it come about that the average density of matter in the universe is the same all over, or the temperature of the Cosmic Microwave Background is roughly the same in all directions?

Then there is the observation that the average density of the universe is at what is called the "critical value"—if the density is greater than the critical value the universe will collapse in on itself in time, if the value is less than, or equal to, the critical value the universe will expand for ever. Why should the density be at exactly this value?

These three observations are very different and have solutions which are surprising, and led cosmologists to a new conception of our universe and the introduction of inflation.

9.3.1 The Observational Puzzles

We will now describe the three observations in more detail.

9.3.1.1 The Monopole Problem

In the 1980s Grand Unified Theories were popular in particle physics. In these theories three fundamental forces—the electromagnetic, the weak, and the strong forces—are actually not fundamental, but are supposed to be merged into a new, single force at the very high temperatures such as occur for example in the early universe. As a by-product, the theories predicted that a large number of heavy and stable magnetic monopoles would be formed, and we should expect there to be a large number of magnetic monopoles in the universe. However, experimental particle physicists and astronomers have not detected *any* magnetic monopoles, which was a surprise at the time and gives us our first unexplained observation and puzzle. We are led to expect a large number of magnetic monopoles and there are none—why are there none?

9.3.1.2   The Flatness Problem

We discussed in Sect. 6.4.2.2 that in Einstein's theory of relativity the curvature of spacetime and the matter content of the universe are related. Whereas we discussed this on rather small scales, this is true on all scales and therefore also for the universe as a whole.

The three-dimensional space that is our expanding universe can have three different overall shapes or geometries which are called "open", "flat" and "closed". Since we cannot give a useful representation of all of these three-dimensional spaces, we show two-dimensional analogous geometrical objects in Fig. 9.2. On the left we show the surface of a saddle, or pringle, the analogy of an "open universe". An open universe is infinite, we should think the surface in the figure extending in both directions, and similarly the universe extending to infinity in all three space dimensions.

In the middle of Fig. 9.2 we have the "flat universe", which can be represented by a flat surface in two dimensions. This is what most people including cosmologists think of as "normal" three-dimensional space, and is also referred to as "Euclidean" space, as all postulates of classical geometry hold in this space. A flat universe is also infinite and all three space dimensions extend to infinity.

Finally on the right of Fig. 9.2 we have a "closed universe", the two-dimensional equivalent of which is a sphere. This space is finite—if we travel in a particular direction on the two-dimensional sphere we will return to our

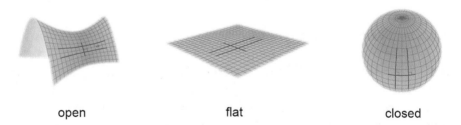

open   flat   closed

Fig. 9.2 The shape of the universe: two-dimensional analogous geometrical objects of three-dimensional space. On the left we have the surface of a saddle, or pringle, the analogy of an "open universe". In the middle we have "flat universe", which can be represented by a flat surface. On the right we have a "closed universe", the two-dimensional equivalent is a sphere. Both open and flat universes are infinite, whereas the closed universe is finite but has no boundary. In red: parallel lines

starting point. This is also true in a closed universe, the three-dimensional equivalent of a sphere: if we travel long enough in a particular direction, we will arrive at our starting point, although in the case of the universe this would take a rather long time. There are however no boundaries in the closed universe, just as there are no boundaries on the surface of the sphere (we are "inside" the surface). We should stress that the universe in all three cases expands (not shown in the figure).

In Fig. 9.2 we also show what happens to parallel lines (red lines in the figure) in each of the three geometries. Here we should recall how we draw parallel lines on a flat surface: we draw a base-line and then draw two lines at right angles to this baseline. This is shown in the middle panel of the figure, a short baseline and the two longer parallel lines. We can extend these lines to infinity if we wanted, and they would remain parallel in the flat geometry.

The behaviour of parallel lines is different in the open and closed geometries. In the open case, we see on the left of Fig. 9.2 that if we follow the procedure described above, the parallel lines do not remain at the same separation, they deviate as we extend them. In the closed case the parallel lines will converge and eventually cross.[5]

The governing equations of Einstein's theory of general relativity show that the average density of the universe is related to the overall geometry. If the average density of the universe has a particular value, the "critical value", the geometry of the universe is flat. If the average density is larger, the universe is closed, and if it is smaller the universe is open. The average density not only determines the geometry it also determines—to an extent—the destiny of the universe. If the density is greater than the critical value the universe will eventually collapse in on itself after some time. If the value is less than, or equal to, the critical value the universe will expand for ever.

When cosmologists measured the average density of the universe it was very close to the critical value, and hence the geometry of the universe is flat or very nearly so. The flatness came as a surprise because it had long been thought that the universe was not flat. New observations which included cold dark matter and dark energy changed the matter content to a value very close to the critical value for flatness. So why is the average density so close to the critical value? Another problem is that the average density will deviate from the critical value

---

[5] We can perform some "geometry in curved space" in two dimensions quite easily. For the closed geometry we can use for example a grapefruit. If we draw an equator on the fruit and follow the procedure to construct parallel lines we will find that they indeed meet at the poles. It is more difficult to find a saddle these days, and we should avoid drawing on it, the same is true for a pringle.

as the universe expands, if it is not precisely equal to the critical value at the beginning, so the universe looks "fine tuned"—specifically set up—to be flat.

9.3.1.3 The Horizon Problem: Why Is the Universe Isotropic and Homogeneous?

Astronomers have found that on a large enough scales, in whichever direction they look and wherever they look in the sky, the distribution of the matter is *on average* the same (see for example Fig. 1.4). Also the Cosmic Microwave Background has the same temperature, 2.73 K, in all directions with only small temperature fluctuations about this value, see Fig. 1.3. Even the distribution of the temperature fluctuation, the distribution of hot and cold spots, is the same in all directions.

This is surprising because we would only expect regions within the causal horizon to be of the same temperature, as inside these regions the temperature can reach equilibrium (the photons can reach all areas inside the causal region and exchange energy until they have the same temperature). Let us illustrate this in Fig. 9.3.

The observer at the centre of the figure measures the temperature of the Cosmic Microwave Background radiation in opposite directions of the sky. Information from both sides took 13.8 billion years to reach the observer, therefore opposite sides of the sky had no possibility to exchange photons or information and reach the same temperature. The observer not only measures the same temperature, also the distribution of small deviations from the average temperature, the pattern of hot and cold spots, is the same.

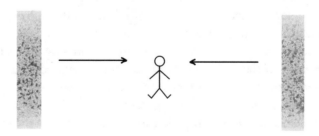

Fig. 9.3 The horizon problem: the Cosmic Microwave Background radiation has the same temperature in all directions, even the distribution of hot and cold regions is similar in all directions. But how can this be, if regions at opposite sides of the sky have never been in contact—it took radiation from the left and from the right 13.8 billion years to reach the observer

Actually the size of a region in causal contact at the time of decoupling is roughly 1° on the sky (see Fig. 8.5 and the discussion in Sect. 8.2.3). We would expect the Cosmic Microwave Background to be isotropic on these angular scales, but not across the sky.

9.3.2 Inflation Provides the Answers to the Cosmological Puzzles

Let us now discuss how inflation solves the questions raised in Sect. 9.3.1. The answers are surprisingly simple.

9.3.2.1 Answering the Monopole Question

We already discussed that the universe expands by 26 orders of magnitude during inflation. The volume increases therefore by a factor of 10^{78}, or 78 orders of magnitude. The number of monopoles created before inflation will be diluted by the expansion of the universe during inflation, the number density of the monopoles drops exponentially so that their abundance drops to an undetectable level—recall, the number density is just the number of particles per volume, if the volume increases by 78 orders of magnitude, the number density decreases by the same order of magnitude. We do not expect to observe a monopole if there are only a few in the observable universe.

This answers the monopole question—the monopoles get inflated away.

9.3.2.2 Answering the Flatness Question

The huge amount of expansion also solves the flatness problem. Imagine living on the surface of a ball with a diameter of, say 100 m.[6] It would be simple to experimentally confirm that you were living on a curved surface in a closed space (just walk long enough in one direction, or draw parallel lines). Now imagine expanding the ball exponentially. As the ball gets bigger and bigger so it would appear flatter and flatter. If the ball were expanded to the scale of, say, the size of the Earth it would appear flat as far as you could see. In this way inflation stretches and flattens any initial curvature of three-dimensional space in the universe.

---

[6]For the analogy to be more precise, if you wish, you could imagine you are two-dimensional living *in* the surface.

But the flat or nearly flat geometry also "drives" the average density to its critical value. And this is indeed the value of the average density we measure today (see Sect. 8.1).

That the underlying geometry of the universe is flat is not only confirmed by measuring the average density of the universe to be equal to the critical density (or at least very close to critical, allowing for observational errors). It was also confirmed by measurements of the distribution of hot and cold spots in the Cosmic Microwave Background as described in Sect. 8.2.3. The photons travelling from the surface of last scattering at the time of decoupling to observers on Earth today will travel on different paths depending on the underlying geometry of the universe. This can be seen with the help of Fig. 9.2: in a flat universe the typical "spot" size is 1°, as discussed in Sect. 8.2.3 and shown in Fig. 8.5. But if the geometry is not flat, the angle under which the spot appears on the sky is either smaller or larger. This is due to the deviation of parallel light-rays (the paths photons travel on) in these non-flat geometries.

9.3.2.3 Answering the Isotropy and Homogeneity Question

The very large amount of expansion also explains how two regions at opposite sides of the universe can have the same temperature, although they should not have had time to exchange information and hence be able to be in thermal equilibrium, that is "agree" on the same temperature. The answer is again simple: the regions appear to have been in thermal equilibrium because they *were* in thermal equilibrium.

As discussed in Sect. 9.1, at the end of inflation the visible universe did fit into a sphere of radius 4.6×10^{-8} lightyears, which is roughly $450,000$ km or about 1.5 lightseconds! The universe had therefore more than enough time to reach thermal equilibrium by the time of decoupling.

As inflation proceeds, any features in the inflaton field get stretched out by the expansion and the field becomes very smooth. If the inflationary epoch lasts long enough for the scale factor to increase by about 10^{26} times then any initial irregularities will be stretched out to lengths that are much larger than the size of observable universe today and are consequently unobservably large: the result is an almost smooth universe with negligible spatial curvature on observable scales—a flat, almost smooth universe with tiny fluctuations in its density.

9.4 An Unexpected Bonus

Inflation was introduced to solve the problems discussed in Sect. 9.3.1. At first this might look like a cop-out, and the reader might rightly say "hang on, every time cosmologists have a problem, they introduce a new particle or field to solve the problem, that's cheating". Instead of resolving the issue, scientists are just moving the original problem to the new problem of explaining where the new particle or field comes from, and nothing would be gained.

If we would solve just a single problem in one particular area of cosmology with a new field, this would indeed be unsatisfactory, and even most cosmologists would rightfully complain, and most of the time the "explanation" would not withstand the test of time. However, if the newly introduced particle or field can answer several open questions, possibly solves a problem that wasn't even thought about when the new particle and field was introduced, then these new particles and fields often become part of the "canon" of physics, or in our case, part of the cosmological standard model. Although we have to introduce new and often strange quantities, whose origin is at first unknown, we end up with fewer open questions than before.

Dark matter was introduced to explain the unexpected galaxy rotation curves, as discussed in Sect. 5.4.1, but then turned out to be essential for structure formation as we saw in Chap. 7. Similarly, inflation and the field driving inflation, named the "inflaton", were introduced to fix the problems discussed in Sect. 9.3.1 above.

But then cosmologists found that inflation also solved the "initial conditions problem", that is how do we get the right distribution of tiny density fluctuations that we need as "seeds" for structure formation, as discussed in Sect. 7.3. These seed fluctuations are needed for the gravitational instability to amplify them, to get the distribution of matter and galaxies on large scales we observe today and to get the right distribution of temperature fluctuations in the Cosmic Microwave Background radiation.

It was this at the time surprising result that convinced most cosmologists that inflation was the correct explanation, despite its weird features—even by cosmology's standards—and the fact that at the time no field of its kind had been found. The only known scalar field in nature, the "Higgs" field, was experimentally confirmed in 2013.

Before we can discuss how inflation generates the initial density fluctuations, we need to introduce "quantum fluctuations". We already briefly encountered Heisenberg's uncertainty relation in Sect. 5.1.5: in quantum mechanics the product of the uncertainty in the energy of a particle and the uncertainty in the time the particle exists at this energy is not zero, but is a very small number. If the time scale of the fluctuation is large, the energy fluctuation is small, but if the time scale of the fluctuation is very short, the energy fluctuation is very large.

This implies that for very short periods energy fluctuations, or particles, can fluctuate into existence and then vanish again. And for extremely small time periods, these quantum fluctuations in the energy density can be very large, by quantum mechanical standards, and become large enough to play a role on macroscopic scales.

Under normal conditions, such as in today's universe, the quantum fluctuations have no observational effects, the fluctuations exist for a very brief moment and then vanish again without leaving a trace in the end. However, during inflation the universe is very different and these fluctuations get carried along and stretched by the very rapid expansion of the universe to very large scales, *before* they can vanish again.

Also, quantum processes are causal processes, which means they can only proceed within a causal region, that is inside the causal or particle horizon. As mentioned previously, a causal region is a region that is small enough to allow for an exchange of light signals or information within it.

Let us now discuss how inflation generates the initial density fluctuations. During inflation small quantum fluctuations in the scalar field, which we might envisage as a "soup" of inflaton particles, are stretched by the extremely rapid expansion of spacetime to scales that are much larger than the horizon size at the time. Once the fluctuations are larger than the horizon—are outside of the horizon—they become "frozen in", they become stuck and stop fluctuating. Because quantum processes are causal, once outside of the causal region the fluctuations can no longer pop out of existence. The scales or distances over which these fluctuations are stretched are much larger than the horizon—the distance light, or anything else, could have travelled in the time since the beginning of the universe. The fluctuations have been stretched to super-horizon scales.[7]

These density fluctuations on super-horizon scales are a unique feature of inflation. They are essential as initial conditions for the formation of large scale structure in the universe. Inflation sets the initial conditions and starts the "clock" for the following evolution of the universe, such that the structure formation process begins everywhere in the observable universe at the same time—at the time inflation ends.

---

[7]Inflation does not violate causality: information exchange takes place in a causal region *before* inflation. Inflation can't be used to transmit signals faster than light.

The density fluctuations need to be of the "right" size and amplitude to get the observed distribution of hot and cold spots in the Cosmic Microwave Background and the observed large scale distribution of galaxies, and clusters of galaxies in the universe today. Inflation imprints the right size-distribution on the density fluctuation for structure formation later on,[8] and also provides the right amplitude for these fluctuations, 1 part in 100,000 or 10^{-5} relative to the average density, on super-horizon scales.

The density fluctuations also provide us with information on this early epoch of the universe, as different models of inflation will give slightly different density fluctuation distributions. The different distributions have observational consequences, as the initial conditions set during inflation are "processed" through structure formation and can then be measured in distribution of hot and cold spots the Cosmic Microwave Background and the distribution of galaxies, and clusters of galaxies on large scales in the universe today.

After the end of inflation the scalar field decays into radiation, normal and dark matter. The decay leaves the amplitude and distribution of the fluctuation in the total density unchanged, the quantum fluctuations in the scalar field are inherited by the density fluctuations in the matter fields.

The horizon, the causal region, grows as time passes and the universe expands. While larger or outside of the horizon the small density fluctuations remain constant—frozen in—until the horizon has grown to a similar size as the fluctuations, when they "enter" the horizon. Once inside the horizon they act as "seeds", setting the initial conditions for structure formation in the universe. Throughout the following history of the universe these small fluctuations grow and are further amplified by the gravitational instability. Through these processes the patterns, formed during inflation, become imprinted on the Cosmic Microwave Background and the Large Scale Structure, the distribution of matter in the universe on the scales of hundreds of millions of light years and more.

During inflation the density fluctuation can also give rise to gravitational waves. The density fluctuations generated during inflation contain huge amounts of mass which is being moved around and accelerated extremely rapidly, hence giving rise to gravitational waves, as discussed in Sect. 6.4.2.3. If we can detect these gravitational waves, and measure their amplitudes and

---

[8]Inflation provides what is called a nearly "scale-invariant" spectrum of density fluctuations. This means that the size-distribution of the fluctuations does not depend on or change with scale: the size of the fluctuations is proportional to the horizon size, which is roughly constant during inflation.

wavelengths, they will provide us with a direct window on this earliest epoch in the history of the universe.

These gravitational waves leave imprints in the Cosmic Microwave Background, but not in the form of density fluctuations or hot and cold spots, but by polarising the radiation we receive. Unfortunately this effect is even smaller than the temperature fluctuations and has therefore not yet been detected. Microwave observatories, like BICEP discussed in Sect. 3.3.3.1, are expected to detect the polarised microwave signal in the coming years. The gravitational waves could eventually also be detected directly by gravitational wave observatories on Earth or in space, with the next generation of detectors.

The density fluctuations formed during inflation might collapse during radiation domination, before nucleosynthesis, and form black holes, if the fluctuations are large on much smaller scales than those relevant for structure formation. These black holes, referred to as "primordial black holes", are different from the ones that form through the core collapse of a star at the end of its lifetime, and form through the gravitational collapse of large overdensities of the radiation. The fluctuations that form black holes before nucleosynthesis are roughly the size of the horizon (or smaller) at this time, with amplitudes at least 10,000 times larger than the fluctuations responsible for structure formation.[9]

Primordial black holes have different masses compared to the stellar remnants, and are possibly responsible for the observed black hole merger events discussed in Sect. 6.4.2.3. Originally thought of as being a viable dark matter candidate, at the moment it is thought that these black holes—if they exist—only make up a tiny fraction of the dark matter, and the majority of the dark matter is in the form of heavy particles, like WIMPS.

These days most cosmologists take inflation seriously because of the explanations and solutions it provides, as we discussed above. Inflation has provided a number of predictions which have been tested and confirmed, such as the distribution and size of the density fluctuations. Despite its unexpected and very special properties inflation answers more questions than it generates which has made it hard not to include it in standard cosmology.

This chapter on inflation concludes our overview of the "cosmological standard model" as laid out in the previous chapters of this book. As mentioned in Chap. 1 this model represents the overall consensus of the scientific community and is still an area of very active research. The cosmological standard model is

---

[9]The density fluctuations responsible for structure formation are on much larger, super-horizon scales at this time, with amplitudes of roughly 10^{-5} relative to the average density.

also referred to as the "Lambda-CDM" model of cosmology, highlighting the ingredients it needs in addition to normal matter and radiation to explain our observations of the cosmos, that is a cosmological constant—"Lambda"— and cold dark matter. Although not in the name, the Lambda-CDM model also requires inflation.

In the following final chapter we will discuss some of the issues and problems the cosmological standard model has.

10

What Lies Beyond?

In this final chapter we will touch upon some more speculative topics that we only mentioned previously in this book but didn't discuss. We will also highlight some of the shortcomings and problems of the cosmological standard model, before we conclude.

10.1 Beyond in Space

As discussed in Chap. 2, one of the foundations of modern physics is testability: we need to be able to test a model or hypothesis, either by performing an experiment or using observations. If the model passes the test we know we are on the right track, if it does not pass the test, we throw it away or adapt it, and then perform new tests. But if we cannot test a model or theory, we can neither rule out nor "confirm" the model or theory. The scientific method does not work in this case, and we are no longer on sound scientific ground.

In cosmology we most often cannot perform an experiment and therefore have to rely instead on observations, and use the observational data to compare our theoretical predictions with. But what about regions of the universe that we can't access through observations? What can we say about the universe beyond the current visible horizon?

We described in Sect. 6.4.4 that today we can receive information from inside a sphere of roughly 28.4 billion parsecs in diameter. Beyond a distance of 14.2 billion parsecs or 42.6 billion lightyears, the distance to the horizon today, it is therefore difficult to test whether our models of the universe agree

© Springer Nature Switzerland AG 2019
K. A. Malik, D. R. Matravers, *How Cosmologists Explain
the Universe to Friends and Family*, Astronomers' Universe,
https://doi.org/10.1007/978-3-030-32734-7_10

with the observations. But no cosmologist would assume that "this is it", that the universe does not extend further than the horizon! So what does lie beyond?

Inflation helps us to vastly extend the region where it is safe to assume the universe looks similar to the region we can observe, because the observable universe was at the beginning of inflation just a tiny part of a region in thermal equilibrium—and hence being very similar or even identical to its surrounding regions—that then got inflated by at least 26 orders of magnitude as discussed at the end of Sect. 9.1. This makes it extremely unlikely that the universe changes immediately in a significant way beyond the horizon.

But what about regions many "observable boxes", cubes of the size of the visible universe, away from our "box"? There is indeed very little we can say about this.

If we had to speculate we would either assume that the cosmological principle, that the universe everywhere is the same, does indeed apply to the whole universe. The universe would look the same or at least very similar to observers many horizon distances away, as it does look to us. This would be the maximally boring option.

Or the universe changes very gradually if we move further away from our own observable region. This might be due to different initial conditions, leading to a different distribution of hot and cold spots in the Cosmic Microwave Background and a differently appearing cosmic web. The universe would look different in regions very far apart but we would still recognise it as "our" universe everywhere.

But neither option can be tested and is therefore mere speculation. Even more speculative would be to assume that the laws of nature are changeable and could be different in different regions of space. We will return to this point in Sect. 10.1.2.

10.1.1 Problems with Infinity

Although inflation provides us with the opportunity to make an educated guess about what lies beyond our immediate horizon, it also poses another possible problem. We discussed in Sect. 9.3.2.2 that inflation drives the overall geometry to being either flat or at least very close to flat.

Arguably today the most popular geometry for the universe is flat—it fits the observational data and is conceptually simple, which explains why cosmologists like it. However, the problem with this geometry, and with the open geometry, is that it implies the universe is also infinite. What does this mean?

Since infinity is a difficult topic in physics, let us first look at the case of a universe that is only close to flat but is closed. In this case the overall curvature of the universe is very small, but not zero, and the universe is very large but finite. By very large we mean we can travel many times the distance corresponding to the size of the observable universe before we return to our starting point. The universe is very large but still finite, it would also contain only a finite amount of matter (or energy).

However, at the present day the observations indicate the universe is flat, and has overall zero curvature (recall Fig. 9.2). This means that the universe is infinite, if we travel in any direction we could travel forever without returning at our starting point. The universe is infinitely large and contains an infinite amount of matter. Mathematically this is not a problem, for example the average density would still be well defined. But this is conceptually difficult.

For example, if the universe is indeed infinite, it would contain an infinite number of "observable universe" sized boxes. Does that mean that everything that possibly could happen does happen, and infinitely often? This sounds even by cosmology's standards *extremely* weird.

10.1.2   Laws of Physics and the Universe

In this book we assumed that there are physical laws that do not change, either in time or in space. This assumption is at the very core of the scientific method. If the laws of physics would change randomly, we would not be able to test our models in a meaningful way.

We can only make sense of the physical world if the laws of nature are either permanent, or change in such a way that we can understand them nevertheless. If this sounds like a circular argument, the reader would be correct. If the laws of physics change in a predictable way, we could simply include how the laws change as another law. This is all that the scientific method could allow for.

For example, at the moment we assume that the gravitational constant that governs how energy curves spacetime does not change with time or space and is indeed a constant. However, we could also allow this quantity, controlling how the curvature of spacetime couples to the energy or matter content of the universe, to change with time and space, promoting it to a "field". Nothing stops theorists from doing this, and this has already been implemented and is nowadays a viable theory. However, the predictions made by this theory fit the observational data just as well as the standard theory, where the coupling between curvature of spacetime and matter content is a constant; this theory therefore attracts less attention these days.

But we could allow all constants of nature to vary with time or space or both, they would be no longer constants but fields. In this way we could also have very different looking regions of the universe, as different values for these "former-constants" in different regions would lead to different physics, the processes described in this book would unfold differently in regions with different values for these constants than in our region. Hence different, distant regions of the universe would not only look different, they would be governed by very different physics.

We could test this however only if we could show that the constants of nature change in our observable universe. Ideally we would perform an experiment that shows this. But so far the laws of nature seem to be permanent and the constants of nature seem to be indeed constant.

10.2 Beyond in Time

Throughout this book we avoided using the terms "time zero", or "Big Bang" and instead simply talked about the beginning of the universe. We avoided the term "Big Bang" as this is not well defined—some cosmologists use it to mean time zero, others to mean the beginning of the radiation dominated epoch etc.—and we avoided time zero since it does not make much sense mathematically. The earliest time we put numbers on is the beginning of inflation, when the universe was "already" 10^{-36} s old.

But what was before inflation? Most of our colleagues would probably answer this question with the "Planck epoch", which is a postulated epoch were all four fundamental forces unified. We do not know much about this epoch, as there are at the moment no credible and agreed upon theories that would describe this even hotter and denser epoch. Also, the Planck epoch will not take us to time zero, it is supposed to start very shortly before inflation, but not at time zero. We therefore think it is more honest to say, we don't know what was before inflation.

This problem is made worse by one of the features inflation was introduced for: inflation dilutes or inflates away whatever was there before, smoothing out any traces that could give us an idea of what went on before—we can't "see" beyond inflation. Any evidence of previous epochs is not accessible to observations.

If we can't say anything about the very beginning of the universe, can we make any predictions about the future evolution of the universe? This is also difficult. Most cosmologists might be comfortable with the statement that there will be no major changes in the evolution of the universe in the next couple of billion years. But beyond that time, things become more tricky.

Let us assume the dark energy is not a cosmological constant but a scalar field—in this context usually referred to as "quintessence". Since we don't know what the quintessence field is precisely, we can also only speculate how it might behave in future. It might decay into radiation and other particles at some point in the future, like the inflaton—the field driving inflation—or it might not. In which case the late time acceleration of the universe could go on for much, much longer or never stop.

This is similar to identifying the dark energy with a cosmological constant. If the dark energy is just the cosmological constant, then the acceleration of the universe that started 4 billion years ago would also never stop. What would that mean? Eventually the expansion of the universe would be so rapid that also bound structures would be affected by the expansion. At first galaxy clusters would be expanded apart, then galaxies, stars and eventually even atoms. But this is only the case *if* the cosmological constant behaves as we today think it does. This is a big assumption—we don't know much about the cosmological constant—which brings us to the final section of this chapter and book, how well do we understand cosmology?

10.3 Problems of the Cosmological Standard Model and Beyond

We can distinguish between unsolved problems already within the cosmological standard model as described in the previous chapters of the book, and problems outside of the cosmological standard model that will require a change or at least an extension of the model. But this separation and distinction is rather artificial and solving problems in one set might lead to solving all or some problems in the other.

These are by no means complete lists, and reflect more what the authors found at the time of writing most interesting.

10.3.1 Problems of the Cosmological Standard Model

The cosmological standard model that we discussed in the previous chapters of this book has at its foundations general relativity, and quantum mechanics and quantum field theory. General relativity describes the evolution of the universe on large scales, and quantum mechanics and quantum field theory govern the micro-cosmos, the universe on the smallest scales. Quantum field theory in particular allows us to include the extremely successful (well tested and in agreement with experiments) standard model of particle physics in cosmology.

In order for the cosmological standard model to work we need to include, besides general relativity and quantum physics, dark matter, dark energy, and inflation. With these ingredients we can explain most of the observations in cosmology, making the "Lambda-CDM" model the preferred model in the scientific community.

But as discussed previously, the problem is that the ingredients of the cosmological standard model are not well, sometimes not at all, understood: what is the nature of the dark matter and the dark energy, what drives inflation? All direct searches for dark matter, for example in particle physics laboratories, have so far found nothing. The same is true for dark energy: nobody knows what it is, whether it is identical to the cosmological constant, or whether it is a scalar field. The inflaton, the scalar field that drives inflation is also not well understood, all we can say with some confidence is that it is not one of the particles known or predicted in the standard model of particle physics.

10.3.2 Problems Beyond the Cosmological Standard Model

But there are more problems beyond the cosmological standard model. We already discussed that general relativity and quantum mechanics are the foundations of modern physics and cosmology. But at the moment both theories seem to be difficult or impossible to combine and reconcile with each other. Our present formulation of quantum mechanics requires absolute space and time, whereas general relativity, as discussed in Sect. 6.4.2 does away with absolute space and time and treats them as four-dimensional spacetime. In modern cosmology, as already explained, we use quantum mechanics on small scales and then use these results as initial conditions on large scales. However, this is not really satisfactory.

There are also deeper conceptual problems. For example, if the universe is infinite as discussed above, what does that mean? The underlying geometry or shape of the universe, its topology, could also be more complicated than the simple cases we discussed here in this book. But how can we open up these questions to rigorous testing?

These problems are more fundamental than the ones discussed in Sect. 10.3.1 concerning the cosmological standard model. It is therefore possible or even likely that solving one of these more fundamental problems will resolve one or more problems of the cosmological standard model itself.

10.4 Concluding Remarks

Cosmology is both a very young and a very old scientific subject. It has made tremendous progress over the last century, despite the shortcomings touched upon in the previous section. There are big holes in our knowledge, and we are only at the beginning to truly understand the universe. But we think this is actually a reason to be optimistic: we have already made some progress with the limited means we had at our disposal—both in terms of theory and observational data.

In the coming years, new observational instruments and experiments promise to provide an avalanche of new data. This new data will not only be a quantitative improvement it will also be of better quality. We will also see previously unavailable data sources come on-line, such as more gravitational wave observations. The new data becoming available will allow us to test and refine our models, develop our theoretical understanding further, and determine new avenues of research.

One such research direction that has become of increasing importance and popularity in the last decades is numerical cosmology, as discussed in Sect. 2.5 and throughout the book. It allows us not only to simulate the universe using the laws of physics as we understand them, and compare the results to observations. The simulations have become so detailed these days, that cosmologist can study the simulations using similar tools with which they study observational data from, say, a galaxy survey.

Figure 10.1 shows an example of such a simulation. The image shows the distribution of dark matter in the universe today, a simulated "slice" through a box with side-lengths of about 764 million parsec or 2.5 billion lightyears. The slice has a thickness of 24 million parsec or 72 million lightyears. Denser regions are brighter, and we clearly see the large scale structure of the matter distribution, filaments around underdense regions and voids—the cosmic web.

The new feature of this simulation is that it uses Einstein's theory of general relativity to calculate the gravitational forces, whereas most previous simulations used Newtonian gravity, adapted to an expanding universe (including the simulations shown previously in this book).

We hope that this book will not only be of interest to the core target audience we had originally in mind when we started writing it—our friends

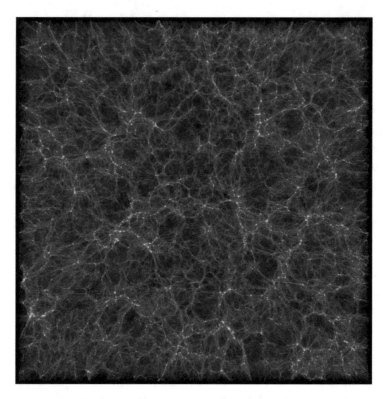

Fig. 10.1 A recent simulation of the universe: the image shows the distribution of dark matter in the universe today, a simulated "slice" through a box with side-lengths of about 764 million parsec. The slice has a thickness of 24 million parsec. Denser regions are brighter. The large scale structure of the matter is clearly visible, filaments around underdense regions and voids—the cosmic web. This simulation used Einstein's theory of general relativity to calculate the gravitational forces, whereas most previous simulations used Newtonian gravity, adapted to an expanding universe (including the ones shown previously in this book). *Image credit: Julian Adamek*

and family. It would be nice if the book would also spark the imagination of younger readers, who might become interested in the subject. Who knows, maybe it will trigger the curiosity of a young scientist who will later on answer some of the many still open questions we touched upon in this book.

Whereof one cannot speak, thereof one must be silent.—Wittgenstein

A

What Didn't Fit: More Technical and Side Issues

In this brief appendix we discuss more technical issues like for example numbers, units and equations.

A.1 Numbers Large and Small

We are all familiar with numbers from our daily lives. We know that an hour has 60 min, a year has 365 days, a table is of the order of a metre high, and we have usually no problems thinking about these numbers.

But we find that even the numbers we operate in every day life can become rather abstract. For example, although a year is a familiar period of time, if we express its length in terms of hours we find that there are 8760 h in a year, expressed in minutes a year is 525,600 min long, and in seconds we find that a year is 31,536,000 s long. These are by no means particularly large numbers, a year has roughly 32 million seconds, but "millions" are already quite removed from our daily experience.

As an example of a number where our experience most likely doesn't help in grasping its meaning let us take the Gross Domestic Product of the European Union in 2018, which was 16 trillion Euros. A trillion is 1,000,000,000,000, a "1" followed by 12 zeroes, which doesn't really help. However, we can express a trillion also as a thousand billions, or a million millions, which helps at least a bit to grasp the size of the number. We can express this number also in

© Springer Nature Switzerland AG 2019
K. A. Malik, D. R. Matravers, *How Cosmologists Explain the Universe to Friends and Family*, Astronomers' Universe,
https://doi.org/10.1007/978-3-030-32734-7

"scientific notation", 16 trillion Euros can be written as 16×10^{12}, the exponent 12 giving the number of zeroes.[1]

Let us construct some illustrative examples:

- Consider a cubic metre, a cube of 1 m side lengths, divided into millimetres contains 1 billion cubic millimetres or 10^9 mm^3,
- a cubic kilometre: a cube of 1 km side length contains 1 billion cubic metres or 10^9 m^3, or 10^{18} mm^3.

Numbers in nature

- a cubic centimetre of water at room temperature contains roughly 3×10^{22} water molecules,
- there are roughly 100 billion, or 10^{11}, stars in our galaxy,
- the size of the visible universe is 92.8 billion lightyears, or 8.78×10^{26} m.

Throughout this book the numbers we have to use are often mind-boggling, either extremely large, like the number of water molecules in a cubic centimetre of water, or extremely small, like the distance between these molecules. This is however sometimes the effect of choosing the wrong units. In the case of the numbers of water molecules, 3×10^{22}, there is nothing we can do, it is just really large.

The distance between the molecules before they bump into each other, the mean free path, is however different. The distance is roughly 10^{-10} m, which seems to be a small number. We can however make it appear even smaller, by choosing different units. Expressed in lightyears, where 1 lightyear is roughly 10^{16} m, the average distance between water molecules is 10^{-26} lightyears. Using the unit lightyear in this context isn't wrong, but wouldn't be a good choice.

On the other hand, we could also choose to use more appropriate units, when we discuss inter-molecular distances. Using "Angstroms" as a unit, where one Angstrom is exactly 10^{-10} m, we would find that the molecules are now separated by single digit, order of unity, distances.

We could therefore introduce special units for each occasion, say in the early universe, then later on if we want to discuss the Cosmic Microwave Background, and so on. But instead of simplifying things, this would make the discussion more complicated, and lessen our ability to understand the

---

[1]We tried to avoid using the "scientific notation" for numbers where possible in the book, writing out numbers instead. However, when this became too clumsy or convoluted, we used scientific notation.

Table A.1 Length units used in astronomy

Unit	Value (m)	Name
1 pc	3.09×10^{16}	parsec
1 kpc	3.09×10^{19}	kiloparsec
1 Mpc	3.09×10^{22}	Megaparsec
1 Gpc	3.09×10^{25}	Gigaparsec
1 lyr	9.4607×10^{15}	lightyear

underlying physics. Hence we are often stuck with, at first glance, ridiculously small or large numbers.

A.2   Units

We have already introduced length units used by astronomers in Chap. 4. In Table A.1 we summarise them, including their abbreviations, although in the main body of the book we try to avoid abbreviations where possible.

It is also useful to keep in mind that 1 pc is roughly 3.26 lightyears.

As mass units we often use a "solar mass", the mass of the Sun. In SI units, the International System of Units (the most commonly used system of measurement), a solar mass is 2×10^{30} kg.

Occasionally we also need volume units adapted to cosmology: $1 \, \text{Mpc}^3$ is a cube with side-lengths 1 Mpc or 1 million parsecs. One cubic Megaparsec therefore consist of 10^{18} cubic parsecs, or 1 million trillion cubic parsecs.

One cubic parsec is a cube with side-lengths 1 pc (or roughly 3.26 lightyears), and in SI units has a volume of $2.95 \times 10^{49} \, \text{m}^3$.

One cubic Megaparsec therefore is in SI units $2.95 \times 10^{67} \, \text{m}^3$.

A.3   Equations …

A lot of people, including surprisingly some scientists, do not like equations. We therefore decided with heavy hearts to "banish" all equations into this appendix.

A.3.1 "Normal" Equations

We can think of equations as questions. For example, simple algebraic equations answer questions like "what number, if divided by two is equal to three", or "what number squared is equal to 4". Often the unknown number is denoted by "x", and we can put this into "Maths" as

$$\frac{x}{2} = 3, \tag{A.1}$$

and

$$x^2 = 4. \tag{A.2}$$

We can either try out all numbers at our disposal to find the answer to these questions, or the solutions to these equations. Or we can use the methods developed over the centuries by Mathematicians. Both, trial and error and mathematical algorithms will give us as answers to the examples above six and two, respectively.

Since we might mention it in the main body of the book, the square root of a number answers the question "what number squared gives me the desired result". For example the square root of nine is three.

A.3.2 Differential Equations

In the examples above we looked at equations that ask questions about numbers. But we can also try and answer questions like "what function satisfies certain conditions when we specify the way or the rate in which the function is changing".

Let us first define the term "function". A simple way of thinking about a function is as a "machine" that takes as input an object, for example a number, then performs an operation on that object and finally gives as output another object. The operation to perform could for example be "divide by 2", or "take the number and multiply it by itself, in other words, square it". We already introduced functions in the previous section on equations, since equations can be used to define functions.

Instead of taking numbers as input for the function, we can also allow for other objects, namely numbers and functions themselves. If our "machine" takes the gradients or the rate-of-change of functions, functions, and numbers as input, we call them "differential equations". They are questions about functions and numbers, not just numbers.

Let us illustrate this with a simple example. We would like to calculate the trajectory of a ball, the setup is similar to that presented in Fig. 7.2. The equations that describe this particular example are taken from Newtonian physics, and can be formulated in general as: the force acting on an object is equal to the mass times the acceleration of the object.

What does this mean? Acceleration is the time rate of change of the velocity, and velocity is the time rate of change of the position of the mass. The time rate of change of the position of the ball is equal to its velocity. Finally the time rate of change of the velocity, the acceleration of the ball, is equal to forces acting on the ball. In our example, the mass is that of the ball (which we approximate as a point mass, so we need only two coordinates or positions to describe the trajectory of the ball, if we keep things as simple as possible).

The forces acting on the ball are gravity, pulling it down, and air resistance, slowing it down. If we neglect air resistance (again to keep things simple), our problem reduces to: we kick the ball, which we take here to mean we give it an initial velocity, and the only force acting on it is gravity. If we take gravity to be a constant force downwards, "all" we need are some simple mathematical tools, that allow us to relate the rate of change of a function to another function.[2]

We then find by trial and error—just trying out functions until they fit the equations—or again using some mathematical tools to solve differential equations the correct function. In this case we find that the function is the quadratic function (here t^2, if t is the time coordinate), the ball moves along a parabola.

But what kind of parabola? This is then determined by the initial conditions, which also need to be specified in order to solve the problem: where and when does the ball start, in what direction and with what velocity do we kick the ball.

Although the above example uses Newtonian physics, also in Einstein's theory of relativity and in quantum mechanics the physical laws allow us to write down differential equation, that can then be solved using mathematical techniques.

A.3.3 Coordinates: Where We Solve Equations

We need coordinates to place points and objects in space and time. They are simply labels that allow us to relate points and objects with each other and with equations, and relate abstract Mathematics with Physics. Also if we want to measure quantities we require coordinates, again to determine the extent of objects and their positions.

---

[2]For example, the rate of change, the "derivative" with respect to t, of t^2 is t, the derivative of t is 1.

In Einstein's theory of relativity there are no preferred coordinate systems, and the laws of physics do not depend on coordinates or coordinate systems. That means we can and must formulate physical laws in such a way that the laws to not change when we change the coordinate systems.

Einstein's theory of general relativity uses quantities that are specifically "designed" or constructed, to allow the theory to be formulated in a coordinate independent way. These quantities are called "tensors", but are way beyond the scope of this book and this appendix.

A.4   Single Telescopes Versus Multiple Telescopes

Recently there have been two developments in observational astronomy: the telescopes that astronomers build and use get larger, and astronomers build telescope arrays. Why do astronomers build larger and larger telescopes? There are two reasons: to increase sensitivity and to improve angular resolution of the telescopes. Let us look at these two terms in turn.

The sensitivity of a telescope is the smallest or faintest signal from a source of electromagnetic radiation—light in the case of a telescope working in the visible wavelength range, radio waves in the case of radio telescopes— a telescope can still detect.The sensitivity of the telescope depends on how much light, how many photons, the telescope can gather. The larger the light collecting area, the fainter the source can be and still be observed. Therefore, the larger the mirror of a telescope or the radio antenna dish, the higher its sensitivity.

The angular resolution also depends on the size of the telescope. By angular resolution we mean the minimum angle or angular separation, under which two separate point light sources can still be distinguished and resolved, that is recognised as two sources, instead of blurring into one.

The angular resolution is inversely proportional to the size of the telescope mirror or radio dish: doubling the dish increase also doubles the resolution. But instead to build a 500 m giant radio telescope, we can also use two small radio-telescope dishes, say with 10 m dishes, and set these two dishes 500 m apart. The resolution of this simple two-dish array will be similar to a 500 m radio dish! However, the two-dish setup's sensitivity will be much worse than the 500 m dish, as a lot less radiation gets "collected" by the two smaller dishes.

How can astronomers use several small telescopes in a telescope array to create a virtual single, large telescope? This is done through "interferometry". The electromagnetic waves received from the individual telescopes are superimposed, which will give interference—constructive and destructive interference as discussed in the context of the LIGO gravitational wave observatory in Sect. 3.3.4.2. The image of the source object that emits the radiation can then be re-created from the many interference patterns using powerful computers.

See the section on the SKA ("Square Kilometre Array"), Sect. 3.4.1 for an example of a radio telescope array.

A.5   Bread Recipe

We include here the recipe for the whole meal bread with seeds used to illustrate the expanding universe in Sect. 6.4.2.4.

We need the following ingredients:

- 500 g whole meal flour,
- 375 g sun flower seeds,
- 100 g unsalted butter,
- 15 g salt,
- 5 g dry yeast,
- 0.5 l water.

All we need to do now is mix the ingredients together, starting with the flour, salt, yeast, the butter and the seeds, then adding the water, until they form a homogeneous mixture. Let the dough rest for an hour, then knead thoroughly. Let it rest for another half hour or so, then put the dough into a baking form, put in the preheated oven (gas level 8 or 220 °C) for about half an hour to 45 min (depending on how dark the bread is supposed to become). Alternatively put all ingredients into a bread maker, and choose one of the whole meal programmes.

Enjoy.

A.6   Further Reading

There are too many good introductory books on the various topics covered in these pages, to list them all. Here we just give a non-representative selection as further suggested reading, mainly books the authors enjoyed themselves.

- On general relativity: Einstein wrote excellent and very readable books explaining his theories, for example

"Relativity: The Special and the General Theory".
- On the scientific method, two classics have recently been republished:
  Thomas Kuhn "The Structure of Scientific Revolutions",
  Karl Popper "The Logic of Scientific Discovery".
- On cosmology: the very readable book by Steven Weinberg "The First 3 min: A Modern View Of The Origin Of The Universe", gives a much more detailed account of nucleosynthesis than we do.

  For the lay person with some mathematics background: Andrew Liddle "An Introduction to Modern Cosmology".

There are also many online resources available, for example the ESA and NASA websites have lots of introductory level material on astronomy and cosmology. Another very useful online resource is WIKIPEDIA.

Glossary

This glossary is intended to define some technical terms as used in cosmology and in the book.

Accelerated expansion The speed with which the universe expands is increasing. One explanation is dark energy.

Anisotropy Deviations from isotropy (independence from directions). Often used in the context of the Cosmic Microwave Background, which is smooth and isotropic with only tiny fluctuations in the temperature of the order of 1 in 100,000.

Antiparticle Each fundamental particle has an antiparticle with the same mass, lifetime and spin (spin is the intrinsic angular momentum), but opposite charge (and potentially other internal quantum numbers). For example, the antiparticle of the electron, the positron, has a positive charge compared to the negative charge of the electron, but is otherwise identical. If a particle and its antiparticle meet, they annihilate into photons (or radiation).

Baryonic matter Baryonic matter is for cosmologists synonymous for "normal" matter, things like for example protons, neutrons, and electrons, and all things made from them, like people, planets, and stars. For particle physicists baryons are particles made up from three quarks, electrons are not baryons in this stricter sense.

Black body An ideal black body absorbs radiation of all wavelengths (whereas a real body always reflects some radiation at some wavelength). It also emits radiation with a particular spectrum (spectrum: the amount of power radiated at a given wave length). The shape of the spectrum depends only on the temperature of the black body and not on for example its material composition.

© Springer Nature Switzerland AG 2019
K. A. Malik, D. R. Matravers, *How Cosmologists Explain the Universe to Friends and Family*, Astronomers' Universe,
https://doi.org/10.1007/978-3-030-32734-7

A black body is very specific and is difficult to reproduce in the laboratory. The black body spectrum of Cosmic Microwave Background is one of closest to an ideal black body spectrum measured in nature.

A black body emits radiation over a range of wavelengths, but there is a "peak" in the emitted spectrum: at this wavelength most of the energy is emitted. The position of the peak depends on the temperature of the black body: at 3000 K (at decoupling) the peak is in the near-infrared wavelengths, at 3 K the peak is in the microwave wavelengths.

Black hole An extremely massive object whose gravity is so strong that nothing can escape its gravitational pull once close (the escape velocity is larger than speed of light), or a region of spacetime from which no radiation (or matter and information) can escape. In the rubber sheet analogy used in Chap. 6, the "dent" made by the object would be replaced by an infinitely deep hole.

We can distinguish different types, depending on how they form and their masses: stellar or "normal" black holes form from stars that collapse at the end of their lives, they have roughly tens of solar masses; primordial black holes form in the early universe from large density fluctuations, are still rather speculative, and come in a very wide mass range, from grams to hundreds of solar masses; supermassive black holes are thought to be at the centre of most galaxies, with masses millions to tens of billions solar masses.

Causal region A region that is small enough to allow all its points to be in causal contact, that is exchange information with each other. Since the speed of light is the maximum speed with which information can be exchanged, the causal region is roughly a sphere whose radius is the distance light can have travelled in the time elapsed (usually since the beginning of the universe).

Comoving coordinates Physical length scales expand with the expansion of the universe. We can factor out the overall expansion of the universe by dividing physical length scales by the scale factor. When we divide physical coordinates by the scale factor, we have comoving coordinates (coordinates that "comove" with the expansion of the universe).

Cosmic web Galaxies are not randomly distributed, but on scales of hundreds of millions of lightyears form structures resembling a web, see Fig. 4.9. The filaments consist of galaxies, gas, and dark matter. Since these structures are separated by large, nearly empty spaces, called "voids", of the order of several 100 million lightyears in size, these structure are also reminiscent of a sponge or foam.

Cosmological principle All locations and all directions, on very large scales, are equal. The universe is homogeneous and isotropic (no preferred positions or directions). The universe is smooth. This is in agreement with observations, we observe no features in the distribution of galaxies beyond roughly several hundred million lightyears, the Cosmic Microwave Background is also smooth, with only tiny fluctuations in the temperature of the order of 1 in 100,000.

Dark energy A form of matter that gives rise to negative pressure, which leads to the accelerated expansion of the universe. All other matter constituents (baryonic matter, dark matter, radiation) lead to a slowing down of the expansion (due to their gravitational attraction).

A popular candidate is the "cosmological constant", as originally introduced by Albert Einstein. He also introduced cosmological constant to counteract the gravitational attraction due to normal matter (although he was interested in a static universe).

Dark matter   A form of matter which does not interact electromagnetically, therefore it does not emit light (hence the name "dark"). Dark matter is mainly noticeable due to its gravitational effect (it might interact weakly with normal matter). A popular candidate for dark matter are WIMPS ("Weakly interacting massive particles") particles that interact weakly with normal matter, with masses of the order of the proton mass (or higher).

Electromagnetic spectrum   The electromagnetic spectrum is the totality of all electromagnetic radiation per wavelength. Often we are interested in the spectrum of a particular source of electromagnetic radiation, then we mean the amount of energy the per second or the intensity of the radiation the source emits at a given wavelength.

Energy   Energy is the ability to do work. There are different types of energy: for example potential energy, kinetic energy, heat energy, and chemical energy. Energy cannot be generated, it can only transformed from one form into another. Another way of saying this is: energy is conserved.

Energy and mass are related, the energy is proportional to mass, but the constant of proportionality is the square of the speed of light (a huge number).

A very fundamental quantity in physics: everything that is has energy. Even objects that have no rest-mass energy, like the photon, and are therefore massless, do possess energy.

Energy density   The amount of energy per volume. In cosmology often used synonymously for mass density, as mass and energy are closely related (see above).

Expansion of the universe   Observations indicate that on the very largest scales the universe, all of space, expands. The expansion of the universe is modelled by a scale factor that increases with time.

Galaxy   Stars, gas, and dark matter, held together by their mutual gravitational attraction. Galaxies contain from just tens of thousands of stars in dwarf galaxies, up to a hundred trillion, or 10^{14} stars in giant galaxies. Their size ranges from 1000 to 100,000 parsecs in diameter. The dark matter dominates the mass of the galaxy, the stars and the gas only contribute 5–10% of its overall mass.

Geometry of spacetime   Local geometry: in general relativity energy, and therefore massive bodies, curve four dimensional spacetime. Objects follow the shortest possible path in this curved spacetime.

Global geometry: the overall shape of the universe. For standard cosmology we only need to consider the three spatial dimensions (treating time *here* independently). There are then three possible global geometries, called flat, closed and open. The analogies in two dimensions for these are: the flat geometry, as the name says, not curved, like the surface of a table; it is infinite. The closed geometry is (positively) curved, like the surface of a sphere, it therefore is also finite but without boundary. The open geometry is (negatively) curved, like the surface of a saddle, or a pringle; it is infinite.

Gravitational instability Gravity is attractive, hence in a smooth distribution of matter, for example some gas, if there is a tiny overdensity, this small overdensity will attract more matter and therefore gain more mass. This leads to an even larger attractive force (stronger spacetime curvature), and attract even more matter, leading to a runaway effect, attracting more and more mass. The gravitational instability leads through the growth, and therefore amplification, of tiny density fluctuations generated in the early universe on very small scales to the distribution of galaxies and clusters of galaxies on the largest scales. It is also responsible for the formation of galaxies through the collapse of giant clouds or blobs of dark matter and hydrogen, and the formation of individual stars and planets.

Halo The usually spherical cloud of dark matter surrounding galaxies, and clusters of galaxies. Galaxies form at the centres of these dark matter halos. The diameter of the dark matter halo is more than ten times larger than that of the galaxy it "hosts". The mass of the halo is roughly 10 times the mass of the visible component of the galaxy, namely the stars (see Fig. 5.5).

Horizon and particle horizon In cosmology "horizon" refers to the region beyond which no information can be exchanged, that is the region beyond we can not "see", as in the everyday concept of "horizon". Cosmologists distinguish different, but closely related types of horizon:

1. The *particle horizon*, also called the *cosmological horizon* or the *light horizon*, is the maximum distance from which particles, including photons, could have travelled to the observer since the beginning of the universe. It forms a boundary between the observable and the unobservable regions of the universe.
2. An *event horizon* is a boundary in spacetime from the inside of which events cannot affect an outside observer. The gravitational attraction is so great that escape from inside is impossible. The most common case of an event horizon is that surrounding a black hole. Nothing, including light, can escape from within the event horizon and reach an outside observer.

Note that the *causal region* is closely related to the concept of a horizon: the causal region at a certain time has to be inside of the particle horizon (otherwise information could not be exchanged between points of the causal region). Both the size of the causal region and the particle horizon are *not* not constant, they grow with time.

Inflation A period of accelerated expansion at the very beginning of the universe, roughly 10^{-43} s (that is 43 zeroes after the comma!) after the "start". During this epoch the universe expanded by many orders of magnitude, at least 10^{45} times, and very rapidly so the whole inflationary period is over when the universe is 10^{-33} s old.

Interferometry Using an interferometer to study the properties of waves. In particular, using the fact that waves interfere constructively and destructively, that is, super-imposing waves (of the same wavelength). The amplitude of the waves will add up if the waves are in phase, the amplitude adds up to zero if the waves are out of phase by half a wavelength.

Jeans length The Jeans length is a threshold scale for gravitational collapse: regions larger in size than the Jeans length will collapse, regions smaller than this scale "bounce back" or oscillate in size, instead of collapsing. For regions larger than the Jeans length scale, the gravitational force due to the material contained in the region can not be balanced by the pressure forces acting in the region. The Jeans length is directly proportional to the speed of sound of the region, and inversely proportional to the square root of the density of the region. It also depends on the expansion of the universe (the Jeans length is also proportional to the scale factor).

Nucleosynthesis or light element nucleosynthesis Took place at about 3 min after the beginning when the temperature of the universe had fallen to about 1 billion kelvin. At this temperature protons and neutrons were able to interact and light element nucleosynthesis could take place, in other words, the protons and neutrons were able to combine to form stable nuclei for light atoms, e.g. Helium, Deuterium and Lithium. Once this interaction stopped the ratio of Helium to Hydrogen was approximately 25%. The heavier elements, for example Carbon and Oxygen, were formed in supernova explosions, later on in the history of the universe.

Order of magnitude To give a rough idea of how large a quantity is compared to another, using powers of 10, for example an order of magnitude larger means about ten times larger, six orders of magnitude a million times. It is the logarithm to base 10 of a number, rounded up or down.

Photometry The measurement of the amount of light or more general electromagnetic radiation a source is emitting. Photometry seeks to answer the question "how bright is a source" by counting the photons the source emits.

Quasar Extremely bright object, believed to be a distant galaxy with a super-massive black hole at its centre. Matter falls towards the black hole and heats up during the infall through friction. This is due to the rotation speed of matter orbiting around an object like a black hole: matter moves not like a rigid body (as would a record on a record player), with the rotation speed proportional to the distance of the object to the centre. The rotation speed is instead governed by Kepler's law, and is proportional to the inverse of the square root of the distance. Matter on different orbits around the black hole therefore rubs against each other. As the matter heats up to extreme temperatures, it emits huge amounts of radiation.

Redshift The stretching of the wavelength of light (and other electromagnetic radiation), compared to the wavelength measured in the laboratory. It is called "redshift" as the wavelength of the radiation is stretched, therefore shifted to longer wavelengths (red is at the long wavelength range of the visible spectrum).

The overall effect can have several contributions, the important ones for us here are relative movement of the source of the radiation and the observer (receding from each other), and the expansion of space. Also important in cosmology is the effect of gravitational fields, which also adds to the redshift of the radiation.

The redshift of a source of radiation is defined as the difference between the observed and the emitted wavelength divided by the emitted wave length. The redshift of a source at the position of the observer is by this definition zero.

Scale factor The scale factor describes how the distance between points in the universe change solely due to the expansion of the universe. The scale factor is a function of time only, which implies it is the same at every point in the universe. The expansion is the same everywhere. The scale factor is chosen to be 1 today, and was smaller in the past, since the universe expands. The scale factor is inversely proportional to redshift.

Spectrometry Measuring the spectrum of light (or more general electromagnetic radiation) a source is emitting, that is the distribution of the emitted light per wavelength. At what wavelengths does the source emit radiation.

Standard model of particle physics The theoretical model describing and explaining the properties and interactions of the known subatomic particles. The model is very well supported by the experimental evidence, for example measurements at the Large Hadron Collider at CERN, the European particle physics research facility in Geneva. It is however not complete, as for example it doesn't explain dark matter.

Star By star we usually mean a sphere of mainly hydrogen and helium, which is held together by its own gravity. The pressure and the temperature at its centre is so large that hydrogen fusion starts (hydrogen "burns" to helium through nuclear fusion), and large amounts of energy are released, mainly in form of electromagnetic radiation. A typical example for a "normal" star is our Sun. But there are also other types of star.

Once the star has used up its supply of hydrogen in its core region, it will enter a new stage in its evolution. Most stars will turn into red giants, and then end their lives in a giant explosion, a supernova. Depending on their mass, the stellar remnant after this explosion is a neutron star, a white dwarf or a black hole (if the star is more massive than 1.5 solar masses, it will form a black hole).

Theory of relativity Physical theory about the structure of the universe, space and time.

Special relativity: deals with inertial or non-accelerated physical systems, space and time no longer separate but form spacetime to account for the speed of light being the same for all observers.

General relativity: extends special relativity to allow the study of non-inertial or accelerated systems. Replaces gravity by the curvature of spacetime.

Work In physics, work is the product of a force times the distance along which the force is acting. For example, if we push a book across a table we have to exert a force over that distance to overcome the friction. We have done work.

Printed in the United States
By Bookmasters